Why Does Anything Exist?

Why Does Anything Exist?

An Introductory Exploration

Stephen A. Simon

broadview press

BROADVIEW PRESS
Peterborough, Ontario, Canada

Founded in 1985, Broadview Press is a fully independent academic publishing house owned by approximately twenty-five shareholders—almost all of whom are either Broadview employees or Broadview authors. Broadview is supported by a collaboration with Trent University, a liberal arts university located in Peterborough, Ontario—the city where Broadview was founded and continues to operate. Broadview is committed to environmentally responsible publishing and fair business practices.

Library and Archives Canada Cataloguing in Publication

Title: Why does anything exist? : an introductory exploration / Stephen A. Simon.
Names: Simon, Stephen A., 1966- author
Description: Includes bibliographical references and index.
Identifiers: Canadiana (print) 20250254743 | Canadiana (ebook) 20250254956 | ISBN 9781554817153 (softcover) | ISBN 9781037700330 (PDF) | ISBN 9781460409428 (EPUB)
Subjects: LCSH: Ontology.
Classification: LCC BD311 .S56 2025 | DDC 111—dc23

Broadview Press handles its own distribution in Canada and the United States:
PO Box 1243, Peterborough, Ontario K9J 7H5, Canada
555 Riverwalk Parkway, Tonawanda, NY 14150, USA
Tel: (705) 482-5915
email: customerservice@broadviewpress.com

Broadview Press books are imported and distributed in the United Kingdom and European Union by:
Gazelle Book Services Ltd.
White Cross Mills, Hightown, Lancaster, Lancashire, LA1 4XS
sales@gazellebookservices.co.uk

European Union – Responsible Person (for official use only):
eucomply OÜ
Pärnu mnt 139b14
11317 Tallinn, Estonia
hello@eucompliancepartner.com
+33757690241

Broadview Press acknowledges the financial support of the Government of Canada for our publishing activities.

Edited by Robert M. Martin
Book Design by Em Dash Design

Broadview Press® is the registered trademark of Broadview Press Inc.

PRINTED IN CANADA

1 2 3 4 5 6 7 8 9 10 25 26 27 28 29 30

For Heather
and Lily,
who fills our universe every day with wonder and joy

Contents

Preface

The question of why anything exists has long held a grip on people's imaginations. The world's splendor inspires wonder, and the mystery of its provenance fascinates many. However, the question can also intimidate. Much of the writing on the topic is either narrow and technical (for those who are already experts) or superficial (for those interested in only a surface-level overview). This book is designed to be both probing and accessible to non-experts. Its aim is to engage substantively with a range of abiding intellectual challenges in a style that is inviting and rewarding regardless of one's training.

I initially developed this project in the course of developing a first-year seminar on "Why Does Anything Exist?" at the University of Richmond, which I have taught many times. The book is meant to fit well with a range of undergraduate courses exploring fundamental philosophical issues. At the same time, both the substance and nontechnical language make it appropriate for any general reader intrigued by the timeless question of why anything exists. Thus, it welcomes readers of widely varying backgrounds into discussions that are provocative and richly rewarding.

Working on this book has afforded me the opportunity to have a wealth of conversations with people who have been generous in reading earlier drafts and offering thoughts and suggestions. I am very grateful to my colleagues at the University of Richmond for their comments on portions of an earlier draft that I presented at a meeting of the Ethics Working Group, including Andrew Alwood, Karin Boxer, Kevin Cherry, Richard Dagger, Jess Flanigan, Javier Hidalgo, Colin Kielty, David Lefkowitz, Claudio López-Guerra, Miriam McCormick, Jeppe Platz, Will Reckner, and Nancy Schauber. I also greatly appreciate the many others who read and commented on portions of the work, including Eno Agolli, Amin Amouhadi, Maria Arévalo, Andrew Bassford, Lis Benossi, Thomas Bonn, Gabriel Day, Alexandra da Eira, Bosco Garcia, Martín González Abaúnza, Jeong Ye Eun, Tim Leisz, John Marvin, Wojciech Nawrocki, Eleanor Oser, William Vincent, and Nicholas Wilson. Special thanks are due to James Marks for outstanding research assistance and feedback on drafts of the chapters at various stages. In addition, I am deeply indebted to the anonymous reviewers, and to Archie Fields and Robert M. Martin for invaluable suggestions, as well to Tara Lowes for extremely helpful editing, and to Stephen Latta and Broadview Press for the excellent stewardship of the book. Naturally, none of the people who have enhanced the book bear responsibility for whatever shortcomings remain.

Lastly, this work has benefited greatly from contributions that were less formal but no less meaningful. I am thankful to the University of Richmond students whose engagement, enthusiasm, and discussions in my classes on "Why Does Anything Exist?" have served as a continuous source of feedback and inspiration. I have also tremendously enjoyed and benefited from many fascinating conversations with Howard Kapiloff and Heather Mendelson Simon on topics related to those explored in the book. My deepest appreciation is reserved for Heather, whose support has made this project possible in too many ways to count.

1. Asking the Ultimate Question

The question of why the world exists is a remarkable one. Every day, we ask questions about things going on in our lives, from why a friend is upset to why the air conditioner is making that clanking noise. These questions take for granted that there already is a stage where all of the action happens. The drama of our lives plays out in a world so enveloping that we rarely pause to contemplate why there is one in the first place. Because the world's existence underlies so much of our experience, seeking a reason for it poses a query of the most fundamental sort. It puts us on a quest for ultimate understanding.

The first thing to do is settle on a way of framing the inquiry. That is, we want to be clear about what the question is. There are many situations in which the particular framing of the question might not be especially pressing. In the midst of experiencing the Grand Canyon or another natural marvel, we might have an intense emotional response of awe that something so breathtaking is here at all. In that context, it may be only the immediate feeling of the moment that really matters. However, once we set out on a quest for possible answers, it becomes important to think about what we are asking.

The question clearly has something to do with why things exist. In speaking of existence, though, we may immediately begin to wonder whether we have in mind a particular kind of existence. When we talk about kinds of existence, we are not merely noting that some things are different from others. It is easy to tell that an office chair is not the same thing as a spoonful of balsamic vinaigrette. Nevertheless, a piece of furniture does not have a fundamentally different kind of existence from a salad dressing. What they share in the most general sense is that they exist as *physical* things. That is, they occupy space, persist over time, and causally influence other things in space and time. Consider a billiard ball. The ball occupies a certain region of space and persists over time. When this billiard ball collides into another, it causally influences the movement of the second ball. Any individual physical thing may have different characteristics from another: a dollar bill is green and rectangular, while a dandelion flower is yellow and round. Even so, the bill and the flower share the same kinds of properties: they both have a color and shape, and, more generally, they both exist as objects in the physical world.

While most people take for granted that there are physical things, questions about kinds of existence arise because many people believe that some *non-physical* things exist as well. For example, even if the four freshly baked cupcakes on the countertop are physical things, one might think that the experiences of seeing, smelling, touching, and tasting them are not. Some philosophers believe that these experiences are purely mental things that really exist, even though they do not have a mass or occupy a particular location in space the way that physical things do. Similarly, since there are four cupcakes that we are experiencing, one might think that the number four exists. Unlike the cupcakes themselves, however, the number four does not seem to be a physical thing. The list goes on; for instance, many people believe that one or more gods or other supernatural beings really exist even though they do not have a physical presence.

The particular way that we will frame our exploration is to ask why there is a *physical* world. One reason to center on physical existence is that seeking an explanation for the physical world resonates with the way that questions about existence often get a hold on us. There is a special kind of amazement prompted by the physical world. It is sometimes the scale and exquisiteness of things that spark our fascination. In addition, while it is highly controversial which kinds of non-physical things exist, if any, it is often simply assumed that we are surrounded by physical things.[1]

Although fascination with the physical world will be our point of embarkation, the discussion will immediately and inevitably extend into myriad other areas that have long inspired reflection and debate. After all, many people believe that the world's origin lies in something that is not

1 To be sure, not everyone accepts the existence of physical things. For example, some thinkers maintain that all existence is actually mental or spiritual in character. See Ramesh Chandra Pradhan, *Metaphysical Idealism, a Contemporary Perspective* (Cambridge Scholars, 2023), and George Berkeley, *A Treatise concerning the Principles of Human Knowledge*, ed. Kenneth Winkler (Hackett, 1982). Regardless, a theory seeking to explain the world's ultimate nature must in some way account for the common experience of our world as consisting of tangible objects—apparently physical things—that surround us.

itself a physical object, such as God or the timeless truths of mathematics. Responses like that prompt us to consider the nature of God, mathematics, or whatever is supposed to explain the world's existence. The inquiry will also lead us into overarching questions such as whether we should expect all our questions to have answers, and what makes one answer better than another. Thus, asking why the world exists provokes an array of gripping issues while also showing how they are all interrelated parts of a common conversation. Regardless of anyone's particular views on a range of profound questions, we are *all* immersed in the spectacle of the countless varied things that surround us.

Asking why the world exists poses an unusual challenge. When addressing why-questions, we normally at least know where to look for an answer. For example, we typically assume that explaining why thunder follows lightning calls for scientific investigation. A question about numerical or geometrical relationships naturally turns our attention to mathematics. And we expect an explanation of why a movie was great to discuss things that we value, such as beautiful visuals or compelling themes. However, when it comes to explaining the physical world's existence, it is far from obvious where to begin. Here, we are not just seeking a particular theory, but asking more generally what kind of explanation is the right tool for the job.

Although the question of why there is a physical world—or the "puzzle of existence," as we will sometimes call it[2]—is daunting, it does not leave us completely at sea. We can orient ourselves by recognizing four sources of explanation that are familiar in our daily lives: science; mathematics; value (as when a person's action is explained by the pursuit of a valued goal); and intentional action (as when the reason my arm goes up is that I choose to raise it). A natural starting point is to assess whether any of these explanatory strategies is a good fit for our task. We will also examine a fifth kind of explanation—based in certain foundational laws or principles—that is employed by some thinkers who consider the four more familiar kinds of explanation ill-suited for explaining the world's existence.

The next chapter places our inquiry in historical context by recognizing ways in which it evokes ancient debates while also raising a distinctively modern problem. We then examine five competing approaches to explaining the world's existence, devoting a separate chapter to each one. Chapter Three ("The Scientific Universe") discusses the natural laws of science as a means of explaining the universe's origins. Chapter Four ("The Mathematical Universe") discusses the possibility that physical reality finds its ultimate foundations in the necessary principles of mathematics. In

2 The phrase is borrowed from *The Puzzle of Existence: Why Is There Something Rather Than Nothing?*, ed. Tyron Goldschmidt (Routledge, 2013).

Chapter Five ("The Good Universe") we consider theories that locate ultimate explanation in the value of existence; from this standpoint the world exists because it is good that it does. Theistic accounts—holding that a deity's intentional acts were responsible for the world's creation—are the subject of Chapter Six ("The Divine Universe"). And in Chapter Seven ("The Lawful Universe"), we take up theories based in a less familiar kind of explanation that is sometimes called "nomological," from *nomos*, the Greek term for "law." Nomological accounts hold that foundational laws or principles provide the explanation for the physical world's existence.

Each of these five chapters discusses the following elements of the particular approach under consideration:

- features that make the approach both distinctive and potentially appealing;
- specific theories that illustrate how the approach may be employed to explain the world's existence;
- potential difficulties facing the approach;
- and implications of the approach for other big questions, such as whether human beings are capable of grasping the deepest truths and whether we occupy a special place in the cosmos.

Not everyone embraces the puzzle of existence with the same enthusiasm. Indeed, some thinkers have argued that it does not pose a legitimate question, or have advanced philosophies that render the question superfluous. These, too, are important perspectives on our question, which are examined in Chapter Eight ("Is the Question Legitimate?").

Our exploration, then, will encompass a wide range of viewpoints on why the world exists. Whatever position one finds the most compelling, the subject confronts us with the most fundamental questions, and highlights the difficulties involved in making all of our beliefs about the world fit together. We find ourselves faced with wrenching choices, since adopting answers that help to shore up weaknesses in a theory often end up exposing others. Thus, in the concluding chapter ("Embracing the Inquiry"), we reflect on what wrestling with such challenges may bring to light, not only about the puzzle itself but also about our own most firmly held beliefs.

Questions for reflection and discussion

1. What are the most fundamental questions that you have pondered?
2. Why do you think so many people are drawn to the puzzle of why the world exists?
3. In what settings are you most likely to wonder about the most basic kinds of questions?
4. Does thinking about fundamental questions further human survival in any way?
5. What novels, movies, or art works have you seen that engage fundamental questions?

2.

A Puzzle Both Ancient and Modern

While the puzzle of existence is distinctively modern in certain ways, it also engages us in conversations that have been ongoing for thousands of years. The long history of philosophical thought makes clear that human beings are not satisfied merely with knowing enough about our surroundings to ensure survival. It is not enough for us to know how to avoid predators or which bushes have the most berries. We also yearn to make sense of things. We want to understand *why things are*, in the deepest sense. That aspiration connects us with the earliest philosophers who, like us, sought explanations of the most fundamental sort.

Even before the rise of philosophy, epic poets like Homer and Hesiod addressed profound questions about the cosmos and our place in it. Homer's *Iliad* (written around the eighth century BCE) explored the role of fate in our lives, for instance, and Hesiod's *Theogony* (written maybe a century later) detailed the birth and rise of the first gods. Nevertheless, Homer and Hesiod are not generally regarded as philosophers. One reason is that Homer and Hesiod invoked goddesses—the Muses—as sources of knowledge in narrating their tales, suggesting that human efforts were insufficient to gain crucial insights.[1] By contrast, the individual widely regarded as the first philosopher in the West, Thales (624–548 BCE), did not bow to divine whispers, but instead relied on his own powers of reasoning to establish his views.

The task that Thales set for himself was to identify the world's first principle (or *arche* in Greek). Beginning with an assumption that the *arche* was

1 Pietro Pucci, *Hesiod and the Language of Poetry* (Johns Hopkins University Press, 1977), 1.

material in character, Thales settled on water as nature's originating principle. We can imagine how he might have arrived at such a conclusion. Living in Miletus on the western coast of present-day Turkey, it might well have seemed that the world literally rested on water. In addition, water is vital to life while also being a part of many inanimate things. Most significantly, we can witness water changing form—from liquid to gas or solid—all while remaining the same substance.[2]

Within a few decades, two other natives of Miletus—Anaximander (610–546 BCE) and Anaximenes (585–525 BCE)—joined Thales' quest. Like Thales, Anaximenes proposed a particular substance as the *arche*, but he concluded that air was a more promising candidate than water. Maybe air seemed like a good choice, as it pervades the smallest spaces, and is also vital to life. While Anaximenes offered an alternative to Thales' proposal, Anaximander's approach differed in more radical ways. Thales' *arche* had been a particular kind of material stuff, as water has particular qualities that are easily observed and described. You can touch ice, see steam rising from a cooking pot, and submerge yourself in a pool of water. Thus, Thales' account did not ask people to accept the existence of anything beyond their own sense experience. Yet for that very reason, one might doubt whether it could really provide a foundational explanation. After all, water is not the only substance; other substances have their own qualities. How can water explain the nature of the cosmos generally if it is just one of the many things in the world? Perhaps troubled by questions along these lines, Anaximander contended that the first principle could not be a specific physical constituent, like water, but must instead be something indefinite, without limitations, which he referred to as the "boundless."[3]

Greek thinkers in other cities soon offered their own thoughts on fundamental questions about the nature of reality. Many framed their inquiry around a problem that continues to preoccupy philosophers today: the relation between change and continuity. Our experience of the world is dynamic, as the things around us shift and transform. Nothing we can see or touch stays the same forever. Yet the very idea of change implies the opposite. For something to undergo a change, it must, in an important sense, be the same entity both before and after the transformation.[4] To identify a formerly green banana as presently yellow is to recognize the persistence of the banana through this change in color. How, then, should we think about change and continuity as features of reality? Is there anything that never

2 See Edward Hussey, *The Presocratics* (Charles Scribner's Sons, 1972), 19.

3 Stephen White, "Milesian Measures: Time, Space, and Matter," in *The Oxford Handbook of Presocratic Philosophy*, ed. Patricia Curd and Daniel W. Graham (Oxford University Press, 2008), 486–87.

4 Hussey, *The Presocratics*, 27.

changes? If so, how do the things that change relate to those that do not? On the other hand, if everything changes, what does that mean for our ability to form any stable understandings of the world?

Two Presocratics[5]—Heraclitus and Parmenides—are known for their opposed responses to the problem of change and continuity. Writing in enigmatic aphorisms, Heraclitus (535–475 BCE) emphasized the pervasiveness of change. One of the most famous surviving fragments from his corpus, for example, stressed that rivers did not stay the same but were ever flowing. Another spoke of a popular barley drink that had to be continuously stirred. It was the very nature of the river and the drink to fluctuate. We tend to associate a thing's essence or inherent nature with the idea of being immune from change, but Heraclitus pointed out that some things only remain what they are by continuously changing. His work suggested that reality was characterized by a unity of opposites, such as day and night, winter and summer, or war and peace.[6] Conflict and dynamism were woven into the world's fabric.[7]

The views of Parmenides (born around 515 BCE) could not have been more extreme in the opposite direction. He maintained that the world consisted of a single, eternal being that never altered.[8] Change was an illusion. Now, nothing could sound more absurd than denying the reality of change. You need only open your eyes to see things shifting all around you: words pop up on the screen as you type, the Sun rises and sets, and people walk past you on the street. But it did not bother Parmenides that his positions seemed preposterous, for his entire philosophy was the result of rejecting common sense as a guide to truth. Instead of unreliable sensory observations and gut feelings, Parmenides relied on reflection and logical reasoning. To construct his world view, he identified rational principles deemed indubitable and followed the implications wherever they led. If the starting points were correct and all of the steps in the argument from one claim to another were proper, then the conclusions had to be true.

---

5 The term "Presocratics" refers to Greek philosophers working before or around the same time as Socrates, who lived from 470 to 399 BCE.

6 Heraclitus' Fragment 67 (DK 22B67).

7 W.K.C. Guthrie, "Flux and Logos in Heraclitus," in *The Presocratics: A Collection of Critical Essays*, ed. Alexander P.D. Mourelatos (Princeton University Press, 1993), 198–204.

8 The interpretation of Parmenides described here—as maintaining that there exists only one being—is a commonly accepted one. See David Sedley, "Parmenides and Melissus," in *The Cambridge Companion to Early Greek Philosophy*, ed. A.A. Long (Cambridge University Press, 1999), 113–33. That said, the proper way to read Parmenides' works remains a subject of academic debate. According to the influential reading of Patricia Curd, for instance, Parmenides should be understood not as claiming that only one thing exists, but, rather, that each thing that exists can only be one thing (that is, that it can only hold one predicate or property that makes it the thing that it is). Patricia Curd, "Parmenidean Monism," *Phronesis* 36, no. 3 (1991): 242–43.

One of Parmenides' crucial premises concerned a topic that philosophers still wrestle with today: the idea of non-being. People talk all the time about things that are not. Fiction writers speak of things that never existed. In regretful moods, we lament the road *not* taken. Skeptics about the supernatural claim that ghosts do *not* exist. Since we normally assume that our words refer to something, the challenge is to explain what is going on when we use language to claim that an object does not exist. Parmenides' uncompromising response to this problem was responsible for the extreme nature of his views. He declared that it was not possible to speak meaningfully of non-being at all, since it entailed a logical contradiction. We could not even conceive the idea of non-being, because every thought had to be about *something*. Thus, Parmenides ruled out any idea that relied on the illegitimate claim of non-being. This led him to deny the reality of change, because it depended on the idea of something new coming into being that had *not* previously existed. The illegitimacy of non-being also meant that reality had to be one undifferentiated whole, since any views about the world's compound composition would suggest that it contained some part which was *not* the same as another.[9]

While Heraclitus and Parmenides offered directly opposing responses to the problem of change and continuity, an alternative approach proved more popular: seeking to explain the reality of both change and continuity. To that end, Democritus (460–370 BCE) proposed that all objects were composed of atoms: imperceptible, physically indivisible, changeless entities. These atoms—of which there were an infinite number in infinite varieties—accounted for persistence, because they could not be created or destroyed. At the same time, the atoms could explain change, because they moved through an infinite void to combine in varying ways. While physics has developed since the time of the Presocratics, Democritus amazingly anticipated core insights of later science. Without the benefit of microscopes or particle colliders, he relied simply on the implications of general observations, such as the tendency for objects to wear away over time.[10]

Not long after the development of Atomist thought, Plato (428–348 BCE) offered a radically different approach to the problem of change and continuity that proved enormously influential. Intent on accounting for both change and continuity, he could not follow the paths forged by either Heraclitus or Parmenides. Plato also rejected Atomism, because he was unwilling to accept that reality was nothing but bits of matter in motion. Instead, he proposed that there were two fundamentally distinct realms

9 Hussey, *The Presocratics*, 85, 92–95.

10 Hussey, *The Presocratics*, 144–48.

of existence.[11] According to this view, one realm comprises the particular, material things that we perceive with our senses, like street signs and wine jugs. This realm accounts for change, as particulars come into existence at a specific time and do not last forever. Another realm consists of immaterial, unchanging, ideal "forms." While each individual red thing, for example, is a particular, all of the various red things in the world share something in common that makes them red. The form of Redness is this common nature. Having a reality apart from any particular red thing, the form of Redness never came into existence and will never perish. Similarly, we can discuss beauty in general, independent of any particular beautiful things, and Plato thought that the forms helped us to make sense of this. The property shared by all beautiful things is the form of Beauty. The forms, therefore, account for continuity.

Plato figures importantly in our exploration due especially to his view that there is more to reality than what we perceive with our senses. While physical things are inevitably flawed and dependent on the forms for their being, the forms are perfect and exist independently of any material things. In describing a realm of being that has primacy over the individual things around us, Plato's thought opens the door to explanations for the fundamental nature of the cosmos that are immaterial in character.[12]

Many people today remain gripped by mysteries that also preoccupied the earliest philosophers. Nevertheless, despite the threads of continuity running across the millennia, the ancient world did not tend to frame the puzzle of existence in quite the same manner as we do now. The ancients tended to take for granted that there always existed some kind of physical stuff.[13] By contrast, the question that we are asking today takes seriously the possibility of a state of affairs in which nothing physical existed at all.

The tendency in Greek thought to assume that there had always existed some kind of material stuff was evident even before the flowering of ancient Greek philosophy, in the work of the great epic poets. The portion of Hesiod's *Theogony* recounting the world's origin begins with the line, "First

11 Plato developed this notion of a distinct, superior, non-sensible realm of forms in many of his dialogues. Among the most prominent discussions of forms are those presented in the following: *Republic*, *Meno*, *Parmenides*, *Phaedo*, *Phaedrus*, and *Symposium*.

12 To be sure, Plato was not the first thinker who appealed to non-physical cosmic explanations. Most notably, Anaxagoras, who preceded Plato by about half a century, claimed that Mind (or *Nous* in Greek) governed and caused all things (as reported in Plato, *Phaedo*, 97b ff.). Indeed, Socrates found this aspect of Anaxagoras' thought initially promising. Nevertheless, Socrates grew disappointed by Anaxagoras, lamenting that when it came to the explanation of specific phenomena he made no use of Mind after all, but instead invoked physical causes such as air, ether, and water. Socrates rejected such causes and insisted instead on seeking explanations showing how Mind had arranged everything in the universe for the best. This search ultimately led Socrates to a conception of the Forms themselves as explanations.

13 Kenneth Dorter, *Form and Good in Plato's Eleatic Dialogues:* The Parmenides, Theaetetus, Sophist, *and* Statesman (University of California Press, 1994), 21–22.

came the Chasm," thus referring to the font of existence as something like a formless void.[14] Following the introduction of the chasm, the tale recounts the appearance of the Earth and gods. The pivotal transition is not from non-existence to existence but from featurelessness to a state of definition and order.[15] Hesiod never paused to consider the origin of the Chasm.

The Greek philosophers shared the poets' inclination to assume that matter had always existed, based partly in the belief that the idea of something coming from nothing made no sense.[16] While Plato's *Timaeus* discusses the idea of a powerful artisan (known as the demiurge, or craftsman) who modeled the harmonious and beautiful cosmos after the forms, the account makes clear that this figure shaped the world from a pre-existing chaos of undifferentiated material. Thus, however significant the demiurge's role in fashioning the cosmic order, this craftsman was not responsible for creating the material out of which things in the world were composed.[17]

For those steeped in the Judeo-Christian tradition, the Bible provides the most familiar narrative of cosmic origins. After the opening line attests to God's creation of the world, the text describes God's productive work over the course of six days, beginning with light and concluding with humanity. However, contrary to what many assume, the biblical language does not unambiguously portray God as creating the world from sheer nothingness.[18] Many early Christians understood the relevant passages as suggesting that God shaped the Earth from a pre-existing primordial chaos.[19] Indeed, the question of whether God created the world from nothing or pre-existing material emerged as a subject of controversy among early believers. Due largely to its welcome implication of unlimited divine power, the view that God created the world from nothing ultimately prevailed. Unlike ancient Greek thought, then, the Judeo-Christian tradition embraced the idea that physical existence had a starting point. At the same time, this narrative of the world's divine origin emerged in a context where belief in a creator God was treated as an indisputable starting point. Thus, one accepting the biblical account did not necessarily confront absolute nothingness as a genuine possibility.

It was the German Enlightenment thinker, Gottfried Wilhelm Leibniz (1646–1716 CE) who articulated the question in its modern form: "Why is

14 Hesiod, *Theogony*, trans. M.L. West (Oxford University Press, 1988), 6.

15 Jenny Strauss Clay, *Hesiod's Cosmos* (Cambridge University Press, 2003), 15.

16 The idea that it was illogical to think that something could come from nothing—later referred to by the Latin phrase *ex nihilo nil fit*—is discussed further in the next chapter.

17 Plato, *Timaeus*, trans. Peter Kalkavage (Hackett, 2016), 30a. The other towering figure of ancient Greek thought—Plato's student, Aristotle—also maintained that the world had not come into being, but, rather, existed eternally. See, for example, Aristotle, *Physics*, Books I–II, trans. William Charlton (Clarendon Press, 1970), 1.9.

18 Daniel Rynhold, *An Introduction to Medieval Jewish Philosophy* (I.B. Tauris, 2009), 49.

19 George Karamanolis, *The Philosophy of Early Christianity* (Routledge, 2013), 68–69.

there something rather than nothing?" Leibniz saw the fact that anything exists at all as a puzzle crying out for explanation. Because nothingness would have been simpler and easier than any other state of affairs, he considered it particularly surprising that anything existed.[20] By explicitly framing his inquiry around the distinction between existence and non-existence, Leibniz drew attention to the possibility of absolute nothingness. Unlike the ancients who took the eternal existence of primordial matter for granted and asked why the world took the form that it did, Leibniz and other modern thinkers have wanted to know why there was any physical stuff at all.[21]

There is a notable manner, then, in which contemporary thinkers pose the puzzle of existence differently from the ancients. At the same time, ancient thinkers were no less interested in questions about the fundamental nature of reality. Like us, they wondered if there is some kind of hierarchical structure to the world. Is there one thing—or principle, or kind of existence—that is in some sense responsible for everything else? One may formulate the quest for foundational explanations in various ways. Today many of us are drawn above all to asking why anything exists: Why is there a world in the first place? In the next chapter, we embark on our exploration into this awe-inspiring question, beginning with the idea that the renowned methods of scientific investigation might furnish the key to ultimate understanding.

20 Gottfried Wilhelm Leibniz, "Principles of Nature of Grace and of Reason," trans. Robert Latta and George R. Montgomery, in *Discourse on Method and Other Writings*, ed. Peter Loptson (Broadview Press, 2012), §§ 7–8. For Leibniz, the posing of this enigma was only one step in an argument designed to reaffirm the existence of God. (Leibniz's case for the existence of God is discussed at length in Chapter Six.)

21 For example, the philosopher Martin Heidegger (1889–1976) famously held that no question was more fundamental than why there is something rather than nothing. See Heidegger, "What Is Metaphysics?," in *Basic Writings from* Being and Time *to* The Task of Thinking, ed. David Farrell Krell (Harper & Row, 1977), 112.

Questions for reflection and discussion

1. What distinguishes philosophy from art, literature, science, and other intellectual activities?
2. How reasonable was Thales' proposal of water as the *arche* given the available information?
3. What point do you think Heraclitus was making in talking about rivers and barley drinks?
4. Was Parmenides right to trust his logical reasoning more than what he could see and touch?
5. How do you think the ancients were able to develop Atomism without modern technology?
6. What is most appealing in Plato's view of the relation between change and continuity?
7. How does Leibniz's way of expressing the puzzle of existence change the inquiry?

3.

The Scientific Universe

The Power of Science

Science has tremendous power as a means of investigating the natural world. Some consider it not just the most reliable path to truth but the only legitimate grounds for making factual claims.[1] The discipline's power is rooted in its insistence that beliefs are acceptable only if they can be verified through empirical observations. While people vary in how they think about the world, there is something that we all have in common: the ability to perceive physical things in the world around us.[2] By insisting on observational evidence that can be shared with and confirmed by others, science provides objective standards for evaluating claims about the world.

In science, a belief cannot stand merely on the grounds that it seems right. One advancing a claim must offer predictions about the observations that will be made if it is true. The requirement of making predictions is connected with science's openness to correcting errors. Today's views may be called into question by later observations, and we should believe only what is supported by the best available evidence. The emphasis on making falsifiable predictions enables researchers to sort out claims worth retaining from those that would be better left behind.

1 We discuss this possibility further in Chapter Nine.

2 It is true that people sometimes argue over their sensory perceptions. An amusing example was the 2015 brouhaha over whether a photograph posted on Tumblr depicted a dress that was gold and white, or black and blue. See https://en.wikipedia.org/wiki/The_dress. Nevertheless, techniques for making precise measurements can be standardized and replicated.

The importance of predictions is illustrated by one of the most well-known observations in the history of science. No figure epitomized the spirit of the Scientific Revolution more than Isaac Newton (1643–1727). His theory of gravity, published in 1687, came to stand as an emblem of the excellence toward which researchers should aspire. The theory treated gravity as a force of attraction exerted by all physical objects on one another. Over two centuries later, Albert Einstein (1879–1955) published his general theory of relativity, which reconceptualized our understanding of gravity by treating it as a curvature of spacetime rather than as a force exerted by bodies.[3] While physicists quickly recognized that Einstein's general theory provided an elegant and unified solution to long-standing problems in physics, the theory still awaited several crucial experimental tests of its distinctive predictions.

Most famously, Einstein's theory of general relativity predicted that light would be bent as it passed by massive bodies. Since Newton's theory viewed gravity as an attractive force between bodies, it did not entail that light passing by a star would be bent in this manner. However, Einstein's theory was not easy to test, because only an extremely massive body like the Sun can produce a detectable amount of bending, but solar glare normally prevents observations of light passing close to our star. Nevertheless, during a total solar eclipse in May 1919, the astronomer Arthur Eddington was able to record observations confirming the Sun's distortive effect on light approaching Earth from distant stars. Along with other empirical observations, Eddington's measurements persuaded physicists to adopt Einstein's new model.[4]

The potency of science is attested to by an extraordinary record of tangible achievements. During the sixteenth and seventeenth centuries, Newton and other pioneers of the Scientific Revolution—including Nicholas Copernicus, Galileo Galilei, and Johannes Kepler—provided astounding insights with the use of a new methodology. While the study of the natural world in the previous centuries had rested largely on theology and the works of revered philosophers, the new scientists emphasized the importance of empirical observations and the ability to predict physical behavior in developing their theories. The discoveries and insights from their work generated unprecedented progress in understanding the world.

The advancement has been staggering since modern science's development of methods for deciphering the language in which nature records its secrets. Only 350 years separates us from Newton's discovery that falling apples and revolving planets are manifestations of the same underlying phenomenon. James Clerk Maxwell published his seminal work on

3 Sean Carroll, *The Big Picture: On the Origins of Life, Meaning, and the Universe Itself* (Dutton, 2016), 49.

4 Steven Weinberg, *Dreams of a Final Theory* (Pantheon Books, 1992), 91–93.

electromagnetism only 150 years ago, and the transformative theories of quantum mechanics, and special and general relativity were developed early in the last century. Moreover, technologies spawned by scientific developments—like computers, global positioning satellites, and microwave ovens—have become parts of our lives. The success of modern technologies cannot help but enhance our confidence in the mode of inquiry that produced them.

The Big Bang Theory

Science not only boasts an honored method of establishing claims but also offers particular accounts that describe the universe's origins and development. This was not always the case. As late as the first part of the twentieth century, the prevalent view of the cosmos was that of a static and eternal universe comprising one galaxy surrounded by an empty expanse.[5] Einstein's general theory of relativity, published in 1915, suggested the possibility that the universe was expanding, which contradicted the prevailing model. However, determining whether the general theory of relativity actually indicated an expanding universe depended on direct experimental evidence. In 1929, Edwin Hubble (1889–1953) provided that evidence by using a powerful new telescope to show that the universe really is expanding. The new findings also showed that the more distant a galaxy is from our own, the more quickly it moves away from us. This set of observations gave rise to the inference at the heart of the Big Bang Theory: that everything in the universe began expanding from a common origin. Through extrapolation from the data, cosmologists could effectively rewind the cosmic clock, estimating an age for the universe of about 14 billion years.[6]

Nevertheless, Hubble's discovery was not by itself enough to dethrone the longstanding view of the universe as static and eternal. In the late 1940s, three astronomers—Thomas Gold, Fred Hoyle, and Herman Bondi—developed a "Steady-state" model, which sought to explain how the universe could remain the same in essential ways even though it was expanding. The Steady-state model proposed that the universe as a whole undergoes a continuous process of matter creation, which maintains it at a constant density.[7] By the middle of the twentieth century, the Big Bang and Steady-state models were locked in a battle for supremacy.

---

5 Lawrence Krauss, *A Universe from Nothing: Why There Is Something Rather Than Nothing* (Free Press, 2012), 2.

6 Brian Greene, *The Fabric of the Cosmos: Space, Time, and the Texture of Reality* (Vintage Books, 2004), 229.

7 Weinberg, *Dreams of a Final Theory*, 34–35.

The Big Bang Theory's eventual victory was hard-earned through the confirmation of critical predictions, with the coup de grâce delivered by the discovery of the cosmic microwave background radiation (or "CMB"). Scientists had known since the late nineteenth century that all matter emits thermal radiation. According to the Big Bang Theory, extremely high temperatures blocked the formation of atoms in the early universe. As a result, thermal radiation was absorbed by the electrically charged free particles, particularly protons and electrons. This interference with the ability of the radiation to travel made the universe opaque. However, by the time the universe was about 380,000 years old, it had cooled enough for atoms to form, which allowed the light to travel more freely, making the universe visible.[8] Based on these beliefs, Big Bang researchers hypothesized that the Earth, since its formation, has been constantly bombarded by radiation that dates back to the period when the universe was about 380,000 years old.[9]

At first, astronomers had no way to test their predictions about the CMB. One obstacle was the difficulty of canceling out background noise when attempting to detect radiation that reaches Earth from every direction. In one of science's happy accidents, two radio astronomers finally discovered the CMB in 1964 while working on projects for Bell Laboratories that had nothing to do with the Big Bang Theory. The astronomers—Arno Penzias and Robert Wilson—initially thought their equipment was malfunctioning. However, when the incoming radiation proved ineliminable, they realized what was going on: they had tripped into one of the most important cosmological breakthroughs since Hubble's discovery of the expanding universe. The serendipitously observed CMB exhibited just the sort of evenly distributed, extremely low-temperature radiation that the Big Bang Theory predicted.[10] In contrast with the Steady-state model, the Big Bang Theory described a dynamic cosmos, in which the early universe differed in dramatic ways from the world in which we find ourselves. The observational confirmation of the CMB was critical, because it supported the Big Bang Theory's predictions in this respect.

While models of cosmic origins continue to evolve, and none commands a consensus on every detail, we can nevertheless sketch the best-known account—the Big Bang Theory—in its broad outlines. First, contrary to what one might think from its name, the theory does not encompass an origin event. The inception of the universe remains shrouded in mystery. Disclaiming knowledge about a "time zero," the theory picks up the story an

8 Mary-Jane Rubenstein, *Worlds without End: The Many Lives of the Multiverse* (Columbia University Press, 2014), 150.

9 Timothy Ferris, *The Whole Shebang: A State of the Universe(s) Report* (Simon & Schuster, 1997), 32–33.

10 Martin Rees, *Before the Beginning: Our Universe and Others* (Basic Books, 1997), 49–52.

extraordinarily brief interval afterwards with the rapid expansion already underway. Thus, cosmologists describe the universe at particular periods of time after the Big Bang, though they cannot tell us about a first moment.[11] The quantifiable measures that figure in the narrative extend well beyond the powers of our imaginations. At the one second mark, for example, the universe had a temperature of ten billion degrees, exceeding that of the hottest stars. Since atoms cannot survive such high temperatures, the early universe comprised a plasma of subatomic particles, including protons, neutrons, and electrons.[12] After 380,000 years, expansion cooled the universe sufficiently for atoms to persist, and gravity's attractive force led to the formation of stars and galaxies.

One of the most significant revisions to the Big Bang Theory was developed to address a mystery raised by earlier versions: Why is the universe so similar in every direction? Although our immediate environment is hardly uniform—my desk has a printer on one side and a coffee mug on the other—at the immense scale of galaxy clusters, the universe is expanding at the same rate in every direction. The density of matter distribution is also remarkably consistent, as is the temperature of the CMB in every direction in the sky.[13] Scientists considered these findings puzzling, because uniformities of this kind require the contact and interaction of matter to even out the distribution of thermal energy. Yet different regions of the universe today are too far apart to exchange energy in this manner. Nor would there have been sufficient time during the universe's earliest phases for temperatures to equalize.[14]

With this problem in mind, theoretical physicist Alan Guth suggested a radical idea: that a tiny fraction of a second after the Big Bang, the universe underwent an unfathomably rapid expansion, referred to as "cosmic inflation."[15] According to this theory, the universe was initially so small that the matter could interact sufficiently to produce a great deal of homogeneity. The inflation then tremendously expanded the scope of the universe while maintaining the homogeneity that had already been established prior to the inflation. The later emergence of unevenness resulted from random fluctuations at the subatomic level, leading to large-scale irregularities long after inflation.[16]

---

11 Greene, *The Fabric of the Cosmos*, 272. As discussed below, the scientific models break down when applied to contexts approaching the origin point.

12 Paul Davies, *The Goldilocks Enigma: Why Is the Universe Just Right for Life?* (First Mariner Books, 2008), 49–50.

13 Carroll, *The Big Picture*, 50.

14 Greene, *The Fabric of the Cosmos*, 14.

15 Davies, *The Goldilocks Enigma*, 59.

16 Stephen Hawking and Leonard Mlodinow, *The Grand Design* (Bantam Books, 2010), 129–30.

The rise of the Big Bang Theory has transfigured the conversation on existential beginnings. Previously, the only entrées to the subject were theology and philosophy. The research spurred by the discovery of the expanding universe has imbued theories on the universe's birth with scientific respectability, suggesting that peering into the farthest expanses of space enables us to effectively witness the cosmic genesis. Those committed to rigorous empirical methods of inquiry can now theorize about the world's origin story without apology.

Apart from the specifics of the Big Bang Theory—or any other particular account—many people deny that the physical world could have had a beginning. One reason is that the idea of a cosmic inception seems to require that something came from nothing. If there was a time when nothing existed, and then the universe came into existence, it suggests that existence arose from nothing at all. As noted in the last chapter, the earliest philosophers assumed that there had always been some kind of material stuff, because the idea of something coming from nothing made no sense. This principle has often been expressed in the Latin phrase, *ex nihilo nil fit* (meaning "out of nothing, nothing is created," or "nothing comes from nothing"). Thus, we may refer to the challenge that the principle poses for theories of cosmic beginnings as the *ex nihilo* problem.

Despite doubts of this kind about the idea of a cosmic dawn, some scientists believe that advances in physics since the first part of the last century help to address the *ex nihilo* problem. These developments concern the nature of time, matter, and causation. With respect to each, we will note why scientific views for a long time made the *ex nihilo* problem seem insurmountable before discussing how recent developments might help to address it.

The nature of time is relevant because a cosmic beginning suggests that physical things sprung into existence from nothing at some particular moment. People have speculated about time since the earliest philosophers. Two and a half millennia ago, Parmenides adopted the radical view that time did not exist, declaring that change was nothing but an illusion in an unchanging world.[17] Over a century later, Aristotle repudiated Parmenides, affirming the reality of both change and flowing time. In the early modern period, the scientific giants Newton and Leibniz waged a famous debate about time. Leibniz defended a *relational* theory of time. In this view, time does not have an independent existence of its own. It is simply a relation among various events. If there were no particular objects or events, there would be no time. By contrast, Newton's *substantival* view held that time

17 Parmenides' views are discussed in Chapter Two.

exists apart from any objects or events.[18] For Newton, it was as if a cosmic clock kept perfect time regardless of whether there were any physical things. While the two great thinkers sparked a lively controversy, Newton's view ultimately prevailed. This is significant for our discussion because Newton's view underscored the challenge posed by the *ex nihilo* problem. If Newton was right, then the idea of a cosmic beginning would mean that physical things popped into existence from nothing at a particular tick of the universe's absolute clock. And that, of course, is exactly what the *ex nihilo* principle forbids.

Newton's understanding of time reigned supreme until it was dethroned by Einstein's theory of general relativity in the early twentieth century. In contrast with Newton, Einstein viewed time as integrated in a spacetime continuum with matter and energy, which can curve, stretch, and contract it. This ruled out the possibility that (space)time could exist without objects and events occurring within it. In this respect, Einstein's theory was closer to Leibniz's relational time than to Newton's absolute time. Since the theory integrates time with physical things, this means that there never was a time when nothing existed.[19] It follows that the Big Bang was not an event *in* space and time, but, rather, constituted the origination *of* spacetime itself. Time and physical things came into existence together. Many people have an almost irrepressible impulse to ask what led up to the Big Bang. What was there *before* the Big Bang, we want to know. Under the influence of Einstein's revolutionary ideas, contemporary science dismisses such questions as nonsense.[20] Temporal notions can have no application in the absence of physical objects. As odd as it sounds, the integration of time and matter allows us to say both that the universe has a finite age and that there has never been a time when nothing existed.

A second development that some believe helps to address the *ex nihilo* problem concerns the nature of matter. Until recently, scientists assumed that physical entities could not appear from (or disappear into) nothingness. However, according to models developed during the last century, certain elementary particles spontaneously appear and disappear, as when an electron and its antiparticle (known as a positron) appear from nothing before annihilating one another. The discovery of these "virtual particles" was made by researchers working in the branch of physics known as "quantum field theory" (an outgrowth of quantum mechanics), which focuses

18 Benson Mates, *The Philosophy of Leibniz: Metaphysics and Language* (Oxford University Press, 2003), 227–33. See generally H.G. Alexander, *The Leibniz-Clarke Correspondence* (Manchester University Press, 1956).

19 Broadly speaking, the idea that time came into being with the world—and that we consequently cannot speak of a time *before* the world existed—can be traced at least to Augustine's work in the late fourth century CE (although as a Christian theologian, Augustine attributed the Earth's origin to God's Creation rather than scientific theories). See Augustine, *Confessions*, Book XI.

20 Carroll, *The Big Picture*, 199–200; Davies, *The Goldilocks Enigma*, 68–70.

on physical interactions at the tiniest scales of atoms and subatomic particles. The phenomenon of virtual particles is related to the Heisenberg uncertainty principle, which recognizes limits to the information that can be measured for a particle at a given time.[21] For certain pairs of variables—such as position and momentum—the more precisely that we know the status of one variable, the less precisely that we can know about the other. If you know the precise *position* of a particle, you cannot know anything about its *momentum*.[22] The Heisenberg uncertainty principle also applies to another crucially important pair of variables: the energy in a particular field of spacetime and the rate at which the energy is changing. According to the uncertainty principle, it is impossible for the field to definitively have both zero energy and a zero rate of change at the same moment.[23] This means there is always a chance that a particle will spontaneously appear within an empty region of space. Some physicists have taken the phenomenon of virtual particles as suggesting that the *ex nihilo* principle is not a bar to explaining cosmic beginnings after all.[24]

A third feature of contemporary physics that has implications for cosmic origins concerns causation. Causation refers to the relation between events and the prior conditions responsible for their occurrence. We commonly speak of events as being brought about in a definitive manner by what happened before. The garbage pail is on its side because a raccoon tipped it over, and the grass is wet because it rained. Until the early twentieth century, the prevailing scientific beliefs were consistent with this understanding of causation. According to Newton's laws of motion—widely accepted for centuries—actions were brought about uniquely and decisively by what occurred earlier. Events unfolded with the necessity of a mechanical clock. If you knew everything about the physical conditions at a particular time, in principle you could predict all subsequent states of affairs.[25] Like Newton's view of time, his view of causation highlighted the challenge posed by the *ex nihilo* problem. If all events are determined by previous ones, then the idea of a first event makes no sense.

21 The principle is named after Werner Heisenberg, who introduced it during the 1920s.

22 Hawking and Mlodinow, *The Grand Design*, 70–72.

23 A field represents a region of spacetime that is subject to a force; the total amount of energy stored in the field can change, as when particles within the space emit radiation. The salient point for present purposes is the uncertainty principle's supposition that it is not possible to say with 100 per cent probability at any time both that a field has no energy and that its energy is not undergoing any change.

24 As noted in the next section, however, virtual particles do not really count as an instance of getting something from nothing, since a quantum vacuum—with its specific, identifiable, physical properties—cannot accurately be described as "nothing." Rees, *Before the Beginning*, 161.

25 Greene, *The Fabric of the Cosmos*, 78–79.

However, contemporary science rejects the Newtonian picture of causal determinism. According to today's dominant physics theories, occurrences at the quantum scale—that is, those at the tiny scale of atoms and subatomic particles—are genuinely random. The way that events at this scale unfold is not fully determined by what happened before. Significantly, one example of a random occurrence is the appearance of "virtual particles" in an empty region of space (or "vacuum"). As discussed earlier, virtual particles can spontaneously appear where there previously had not been any physical things.[26]

Together, various ideas of contemporary physics like those discussed have suggested the possibility that the universe itself resulted from a random event in which physical entities spontaneously appeared. The first physicist to put the pieces together was Edward Tryon (1940–2019), who, in the early 1970s, hypothesized that the universe began as a random event in a pre-existing quantum vacuum (or empty spacetime).[27] While science had largely avoided discussion of cosmic beginnings until the middle of the last century, many now believe that the discipline provides just the right prism through which to view the subject.

Until recently, researchers assumed that science could not explain the world's origin due to the *ex nihilo* problem. Deeming it common sense that something could not arise from nothing, scientists did not think they could tell us why anything exists. However, discoveries have persuaded some to reconsider. The Big Bang Theory provides a narrative of the universe's birth, and developments in physics offer potential responses to the *ex nihilo* problem.

A scientific approach to the puzzle of existence would promise so much. Scholars today often envision a division of labor that tasks science with empirically investigating nature while leaving grand, metaphysical questions about reality for more speculative fields, like philosophy. For those who hold this view, the question of why anything exists falls outside science's legitimate scope. But what if scientific theories could encompass an account of why there is a world in the first place? This is the ambition of a scientific approach to why the world exists. Its champions hold that they can take on the awe-inspiring puzzle of existence with the same respected methodologies that have led to revolutions in technology and society. When it comes to the ultimate question, they refuse to abandon the field. Given the power of science as an avenue for investigating difficult problems, the

---

26 Ferris, *The Whole Shebang*, 236–37. For one study of differing views held by practitioners on questions like this, see Sujeevan Sivasundaram and Kristian Hvidtfelt Nielsen, "Surveying the Attitudes of Physicists concerning Foundational Issues of Quantum Mechanics," accessed December 8, 2024, https://arxiv.org/pdf/1612.00676.pdf.

27 Arthur Witherall, "The Fundamental Question," *Journal of Philosophical Research* 26 (2001): 72–73.

prospect that it might hold the key to satisfying our wonder about the very fact of existence is powerfully alluring.

Problems in Contemporary Theories

While the ascent of the Big Bang Theory has garnered tremendous attention, a scientific approach to explaining the world's existence encounters substantial challenges. In the next section, we examine a fundamental challenge involving the nature of scientific explanations. This section considers challenges for a scientific approach that concern limitations and discrepancies in prevailing scientific models.

The excitement generated by the Big Bang Theory was based largely on the tantalizing prospect that cosmologists could rewind the universe's home video all the way back to the start. In fact, however, the theory does not even purport to describe a first moment. Instead, the footage begins a tiny fraction of a second into the expansion, when material existence was already a reality. As the physicist Brian Greene writes, the Big Bang Theory "leaves out the bang. It tells us nothing about what banged, why it banged, how it banged, or ... whether it ever really banged at all."[28] In the words of the renowned physicist Steven Weinberg, "there is an embarrassing vagueness about the very beginning."[29]

Worse, the framework on which the Big Bang Theory depends—general relativity—breaks down at a critical point. Rewinding the clock leads us to a point of infinite temperature and density, which physicists regard as an indication that something has gone badly wrong in the theory.[30] The narrative cannot be traced to the beginning and does not explain why the Big Bang occurred. As characterized by another prominent physicist, Sean Carroll, "the Big Bang doesn't actually mark the beginning of our universe; it marks the end of our theoretical understanding."[31]

Since the Big Bang Theory is set within a larger body of cosmological understandings, major discrepancies in the field provide additional reasons for wariness about accepting the theory at face value. One problem is this: researchers today cannot account for more than a small fraction of the universe's matter and energy, which gives rise to troubling anomalies. Based on the amount of matter that cosmologists can observe, the universe's galaxies

28 Greene, *The Fabric of the Cosmos*, 272.

29 Steven Weinberg, *The First Three Minutes: A Modern View of the Origin of the Universe* (Basic Books, 1993), 8.

30 John D. Barrow, *The Book of Nothing: Vacuums, Voids, and the Latest Ideas about the Origins of the Universe* (Pantheon Books, 2000), 274, 295.

31 Carroll, *The Big Picture*, 51.

should be scattered far away from one another, but instead they are clustered together. Closer to home, the stars in our own galaxy are revolving around its center more quickly than they should be. The most plausible explanation is that large rings of undetected matter are exerting gravitational force on them.[32] Otherwise, the Milky Way "would speedily unravel like an exploding flywheel."[33] Another problem is that the amount of known matter in the universe is not sufficient to account for the clumping in the early universe that would have been needed for galaxy formation. For these and other reasons, cosmologists believe that a large portion of the universe's matter exists in a form (called "dark matter") that has thus far resisted discovery. As a result of this sizable discrepancy, we must take seriously the possibility that current methods of modelling gravitational attraction are substantially incorrect.[34]

Likewise, cosmology also faces a mystery of missing energy. As with respect to matter, scientists have reason to believe that the total amount of energy in the universe is dramatically larger than what they have thus far been able to observe. What brought this discrepancy to light was a surprising discovery made during the late 1990s: that the universe's rate of expansion is accelerating.[35] The finding came as a shock, because it had previously been believed that gravitation was acting as a brake on expansion. Just as cosmologists use the term "dark matter" as a placeholder for the unidentified source of the missing mass, they refer to whatever is causing the accelerating expansion as "dark energy." The degree of the error regarding energy is even greater than it is regarding matter. In physicist Paul Davies's words,

> If you add up the dark energy responsible for making the universe accelerate, you find that it actually represents a total mass that is more than all the matter—visible and dark—put together. It seems that dark energy constitutes most of the mass of the universe, yet nobody knows what it is.[36]

There is an even more serious problem lying at the core of contemporary physics: its two most fundamental theories—quantum theory and general relativity—contradict each other. Quantum theory reigns over events at the tiny scale of particles and subatomic particles, while general relativity—the prevailing theory of gravitation—is critical in situations

32 Rees, *Before the Beginning*, 103–04.
33 Davies, *The Goldilocks Enigma*, 117.
34 Rees, *Before the Beginning*, 110.
35 Carroll, *The Big Picture*, 52.
36 Davies, *The Goldilocks Enigma*, 122.

involving large densities and speeds close to that of light. Each framework is supposed to have universal scope, yet the two mathematical models are not fully compatible with one another, and the inconsistency reflects deep conceptual divergences. One discrepancy concerns the nature of causation. We noted that quantum theory recognizes random behavior: events that are not fully determined by what happened earlier. However, the equations of general relativity that physicists use to model gravity do not allow for any indeterminacy. Furthermore, while the energies of quantum particles are quantized—that is, they exist only as multiples of some constant number—masses and energies in general relativity are continuous, meaning that they can exist in any quantity. (To illustrate, consider the difference between hamburger buns, which you can only buy in specified quantities, and ground beef, which a butcher will sell in any amount.) The problem is that attempts to incorporate gravitation into quantum theoretical calculations by quantizing it (that is, supposing that gravitational force exists only as multiples of some constant) yield deeply unacceptable results.[37] In general, the quantitative machinery falls apart if the models are combined.[38]

Since most research does not require the simultaneous use of quantum theory and general relativity, their incompatibility does not usually pose a practical problem. However, some areas of research depend on both frameworks and, thus, the discordance impedes investigation. In particular, researchers must employ both theories when studying conditions of fantastic density where large masses are packed into tiny areas. One such area of research is crucial here: the attempt to outline a chronological account of the universe's earliest moments. Shortly after the Big Bang, all the universe's mass and energy existed within a region so minuscule that it can only be described in terms applicable to the atomic and subatomic realms, making reliance on quantum theory imperative. Yet the tremendous concentration of mass at that time makes general relativity equally indispensable. As a result, we can only hope to discern what occurred with an account that integrates the two frameworks.[39] Underscoring the problem's central significance, Greene writes: "It's not an overstatement to describe this situation as

37 Greene, *The Fabric of the Cosmos*, 16.

38 Rees, *Before the Beginning*, 158. Another manifestation of the inconsistency between the two theories concerns the phenomenon of non-locality. Relativity theory holds that all physical interactions occur locally—that is, through mediating effects that take place in a finite amount of time, at a speed no faster than that of light. However, according to quantum theory, entangled particles affect each other instantaneously over indefinitely large distances. Indeed, physicist John Bell proved that this feature of non-locality is an ineradicable feature of quantum theory. This theorem—named after Bell—has been experimentally confirmed many times. See N. David Mermin, "Is the Moon There When Nobody Looks? Reality and the Quantum Theory," *Physics Today* 4 (1985): 38–47.

39 Ferris, *The Whole Shebang*, 37–38.

a theoretician's nightmare: the absence of mathematical tools with which to analyze a vital realm that lies beyond experimental accessibility."[40]

Attempts to reconcile quantum theory and general relativity have thus far been unavailing. The best-known approach to the problem—"string theory"—posits that the most elementary entity in the world is a string whose different modes of vibration account for the various phenomena in the observable world. Although the strings themselves are one-dimensional, the theory hypothesizes up to ten or eleven hidden dimensions. While many regard string theory as a promising direction for further investigation, it faces major obstacles.[41] Since the strings are one hundred billion times smaller than an atomic nucleus, the strongest particle accelerator remains trillions of times too weak to observe strings. Moreover, current versions of string theory make no currently testable predictions that could be used as a basis for probing their claims.[42] In sum, the incompatibility between quantum theory and general relativity poses a major problem for science's ability to theorize about the universe's origins.

The Circularity Problem

Even if the problems just discussed could be solved, there is a more fundamental challenge for a scientific approach to explaining the world's existence, which concerns the very nature of scientific explanation. Perhaps tomorrow's research will unify general relativity and quantum theory, and resolve other shortcomings in current scientific theories. We cannot know what the future of science holds. Such a happy outcome would dissolve challenges for a scientific approach regarding discrepancies in prevailing models. Nevertheless, the more fundamental challenge asks whether a scientific answer to the puzzle of existence would be inescapably circular. If so, this might suggest that when it comes to addressing the ultimate question, science is simply the wrong tool for the job.

Although many questions about the practice of science remain controversial, we can make general observations about the form that scientific explanations commonly take, especially in physics, the field with the most direct bearing on cosmic origins. Broadly speaking, science investigates the natural world with an eye to presenting concise summaries of regularities in the ways that things behave.[43] Physical theories typically explain an indi-

40 Greene, *The Fabric of the Cosmos*, 15.

41 Rees, *Before the Beginning*, 159.

42 Rees, *Before the Beginning*, 227.

43 John D. Barrow, *New Theories of Everything* (Oxford University Press, 2007), 10–11.

vidual occurrence by showing that it falls under a general rule. Suppose a question is posed as to why some object behaved as it did. An answer may take the following form: the object falls within a class of entities (call it "X") that has been found to behave in the manner observed ("Y") under the specified conditions ("Z"). Thus, any instance of an X behaving in a Y manner in Z conditions can be explained by reference to the rule. For example, the explanation of why a particular pot of water under specified conditions boiled at 100 degrees Celsius (or 212 degrees Fahrenheit) will refer to general rules—arrived at through past observations—according to which water under those specified conditions can be expected and predicted to boil at that temperature.

When we say that an explanation is circular, we mean that it assumes what it is supposed to explain. I cannot demonstrate that everything in a book is true by noting that it says so on page 79. Rather, I must point to a reason outside the text for deeming it credible. Why might one worry that a scientific answer to the puzzle of existence would be inevitably circular?

Consider that scientific theories do not recognize influences on physical behavior that are not themselves physical in nature.[44] The scientific presumption is that explanations speak in the language of physical interactions. If two experiments yield different results, any attempts to account for the divergence by reference to non-physical influences would be dismissed as falling outside the scope of scientific explanation. Thus, scientific inquiry depends on the assumption of physical existence. This methodology serves science extraordinarily well when it is performing its usual task of investigating regularities in physical behavior. However, the same methodology is problematic when explaining why there is a world, because it assumes the very fact of physical existence which it is supposed to explain.

Some have suggested that developments in contemporary physics can help science to overcome the circularity problem. Based on the phenomenon of virtual particles, they propose that the universe arose as a spontaneous, random occurrence in the quantum vacuum. The idea is enticing because the vacuum appears to be nothing, yet it exhibits quantum behavior, including the random appearance of particles. However, the difficulty

44 While there is no universally agreed upon definition of "physical," we are using the term here in a general way that is not meant to take sides on potentially controversial nuances in the term's meaning and application. For example, scientists might disagree over whether particular concepts used in scientific theories—like gravitational fields or the strings discussed in string theory—should be thought of as physical things (rather than, say, merely useful rubrics for studying the phenomena at issue). Nevertheless, there is generally enough underlying agreement for physicists to think of themselves as studying similar things. That there is sufficient common ground for the everyday practice of science is reflected in certain paradigm cases; there is little controversy, for example, that planets, gases, and crystals are appropriate subjects of scientific investigation, while ghosts, demons, supernatural powers, and abstractions—like justice or hope—are not.

with trying to escape the circularity by appeal to the quantum fluctuation hypothesis can be stated simply: a quantum vacuum is not nothing. To the contrary, the hypothesis inescapably relies on the quantum vacuum having specific, identifiable physical properties. At best, the quantum vacuum suggests an entity that existed independently of the Big Bang, but it does not explain why that entity existed in the first place. Indeed, the phenomenon described as virtual particles can only be understood against the backdrop of a larger system. As cosmologist Martin Rees observes, "The physicists' vacuum is a far richer construct than the philosopher's 'nothing': latent in it are all the particles and ... equations of physics."[45]

Science's reliance on physical things to explain phenomena might not be a problem if it could show that physical existence was necessary. That is, even if something non-physical could not explain the physical world, science might still be able to furnish a satisfying solution if it could show that the physical world effectively explained its own existence. This kind of answer would show that by its very nature the world had to exist.

Let us consider why science seems unable to demonstrate the necessity of physical existence. When a scientific theory tells us that objects of type X behave in manner Y under Z conditions, what is its basis for saying so? Generally, it is not that scientists have a deep understanding of why an X could not possibly do otherwise. Rather, it is because they have observed instances of X's behaving in a particular manner, or their best existing theories entail that X behaves that way. To be sure, the general rule describing X's behavior might fit within a more comprehensive theory connected to other general rules that together tell a more complicated story. In this sense, science may explain why an object behaved as it did by referring to more general descriptions of how that and other objects behave. However, within that larger tapestry of scientific theories, questions about why physical things behave according to the particular patterns that they do remain unanswered. Science does not pretend to know exactly why entities behave and interact in the particular ways that its theories suggest. They just do. In effect, the scientific explanation for why a thing behaved as it did on a particular occasion is that things of that kind generally behave that way under similar conditions. For all their power and usefulness, in the end scientific explanations amount to a claim that something is the way it is because things are the way they are.

We are so accustomed to basic patterns of behavior in the things around us that it is easy to think of them as inevitable. Since birth, we have learned to navigate the world in light of these regularities, which easily generates

45 Rees, *Before the Beginning*, 161.

the illusion that we understand why things behave the way they do. In "An Abstract of a Treatise of Human Nature" (1740), the Scottish philosopher David Hume (1711–76) proposes a thought experiment designed to help us shake off our familiarity with the natural laws.[46] He asks us to imagine a human being brought into existence as a fully formed adult. Called "Adam" to evoke the biblical first man, this individual has all the same sensory and intellectual capacities as healthy adults, but lacks something that you and I take for granted: the experience of how things behave. Having just materialized, Adam has never perceived objects interacting with one another. Now, imagine that this person sees the following: two billiard balls rolling toward each other. The question that Hume's thought experiment provokes is this: can Adam anticipate what will happen when the two balls collide? As Hume points out, it seems that he could not. What basis could Adam have for making such a prediction? Not only has he never seen one billiard ball strike another, he has not observed impacts of any kind, and he has no pre-existing theories describing patterns of physical behavior.

The thought experiment prompts reflection on the basis of our own ability to predict what the second billiard ball will do upon being struck by the first. The only reason we know what will happen is that we have seen similar things happen before. Familiarity gives the impression that we understand why billiard balls must behave this way. Yet there is nothing about the situation that tells us purely as a matter of logical necessity what must happen. The story encapsulates the limits of scientific understanding. Science is terrific at making predictions of how things will behave under specified conditions, but its theories do not tell us why physical objects *must* follow the particular patterns of behavior that they do. The reason scientists can forecast events is not that there is only one way the screenplay could have been written, but that they have already seen the movie.

It is tempting to suppose that scientists will one day be able to demonstrate that the regularities of nature could not have been other than they are. Perhaps there is only one set of regularities that would not fall into self-contradiction. While the idea of such an impending discovery is alluring, there is little indication that it will actually happen. Physicists today can envision many different combinations of natural laws that are internally consistent. No indications have yet surfaced that the features of our universe had to be precisely as they are. Instead, these features present themselves to us as a given that scientists describe without understanding the underlying reasons for them. Physicists often acknowledge this in noting

46 Hume published the work anonymously not long after the publication of his *A Treatise of Human Nature*, which was not initially well-received but has since been recognized as a masterpiece.

that what science really does is explain *how* physical things behave, not *why* they behave the way that they do.[47] As Sean Carroll writes:

> Physicists sometimes fantasize about discovering that the laws of physics are somehow unique—that these are the only ones there could possibly have been. That's probably an unrealistic pipe dream. It's not hard to imagine all sorts of different possible ways the laws of physics could have been.... Perhaps the universe could have been a lattice, like a checkerboard, with bits flipping from on to off as time passes in discrete units. Perhaps the sum total of reality could have been a single point, lacking either space or time. Perhaps there could be a universe that had no regularities at all, one where there would be nothing we would recognize as a 'law of physics.'[48]

It is notable that what arguably undermines science as an approach to explaining the world's existence is tied to what makes it so successful within its usual domain: the commitment to rooting all theories in empirical observations of physical behavior. This is a crucial strength when science is addressing questions about how existing physical things behave, but it becomes problematic in the context of asking why there are any physical things in the first place. The discipline prides itself on following the implications of observations wherever they lead, no matter how counterintuitive. When researchers make new observations that contradict prevailing beliefs, they rewrite their theories. Yet this correctability also reflects the limitations of our scientific understanding: we do not know why things do what they do, except that they seem to follow certain behavioral patterns.

Living in the Scientific Universe

Enmeshed in a network of smartphones and Wi-Fi, many of us today take the validity and indispensability of science for granted. However, our question is not about science's value in general. We are interested in the

47 Krauss, *A Universe from Nothing*, 143; Weinberg, *Dreams of a Final Theory*, 25.

48 Carroll, *The Big Picture*, 203. As noted, there is a sense in which science offers answers to "why questions." If one asks why a ball fell from a cliff at the particular speed that it did, it is not, of course, that science has nothing to say. To the contrary, physics can offer theories involving gravitation, friction, and so on, furnishing a scientific explanation. The important point here, though, is that if one pulls the camera back for a wider view and asks why physical things behave according to the full set of patterns that they do, science has no answers. Ultimately, it provides descriptions—quite intricate and complex—of how things can be expected to behave under specified conditions. It cannot, however, tell us why it had to be the case that the patterns of behavior would be just as they are. Carroll expresses this concisely when he notes: "It's not hard to imagine all sorts of different possible ways the laws of physics could have been."

implications of a scientific approach to the puzzle of existence. One may embrace scientific progress while endorsing another kind of explanation for the world's existence. In this section, we consider what might follow from adopting an answer to the question of why there is a physical world that rests exclusively on scientific explanations.

Now, there are as many ways of combining ideas about the world as there are people who reflect on such things. Thus, we should not assume that specific beliefs necessarily follow from the adoption of a scientific approach to the ultimate question. We can nevertheless make general observations about convictions that seem to fit naturally with a scientific explanation for the world's existence.

As we will explore in later chapters, other approaches to explaining the world's existence posit a more fundamental, non-physical, kind of existence that is responsible for the physical world. By contrast, an exclusively scientific approach begins and ends with claims about the behavior of physical things.[49] Relatedly, a scientific approach assumes that the most fundamental truths are those describing the properties of physical objects, from electrons and protons to gas clouds, acidic solutions, and galaxy clusters. Descriptions of how physical things behave lie at the bedrock of explanation.

Human beings have long been fascinated not only by the nature of the universe but also by our place within it. Questions of this kind include the extent to which the most basic truths are open to our understanding and whether we play a special role in the cosmos. An exclusively scientific approach assumes that human beings have the same kind of existence as everything else that we encounter. There is no fundamental distinction between us and the things around us. We are made of the same building blocks and are answerable to the same behavioral regularities. Although we can do different things from hydrogen atoms, granite rocks, or porcupines, this is only because our physical components are arranged in ways that give rise to different patterns of conduct.

Regarding humanity's ability to grasp the basic nature of reality, a scientific approach has both optimistic and pessimistic implications. In an optimistic vein, our essential similarity to everything else puts us in close contact with the world that we want to understand. As physical creatures, we have evolved to capably navigate the environs. Our bodily senses provide access to knowledge about the world, making us well-equipped for observing how things interact. The explosion of scientific advancements in recent

49 Those adopting an exclusively scientific approach can—and often do—recognize the existence of things that are not physical. Nevertheless, their answer to the puzzle of existence is based exclusively in scientific explanations (and, not for example, in a divine being's creative acts).

years reflects our faculty for developing theories describing the behavioral patterns that are evident in the things around us.

On the other hand, there are also reasons why a scientific approach to the puzzle of existence may suggest limitations on our understanding. One reason concerns limitations on what a scientific account of cosmic origins can explain. No matter how much detail a scientific theory provides regarding the universe's development following its earliest moments, it may ultimately fail to explain why anything physical exists in the first place. Since scientific theories speak in the language of interactions between physical things, they cannot explain physical existence by reference to anything that is not itself also physical in nature. Nor do scientific theories explain why physical objects exist by necessity. As noted, we have no indications that science will be able to show why the existence and nature of physical properties could not be other than they are. From a purely scientific standpoint, the world's existence presents itself to us as a brute fact: it just is. Given this brute fact, science can furnish fascinating and practically vital insights about how physical things behave, but it finally rests on an unexplained starting point.[50]

A second reason why a scientific approach may indicate limitations on our understanding concerns the incomprehensibility of quantum theory. As we have seen, quantum theory is crucial to scientific explanations for the world's existence. Edward Tryon pioneered this strategy several decades ago, proposing that the universe began as a spontaneous, random event in a quantum vacuum. Given quantum theory's centrality, difficulties in making sense of it raise questions about our ability to truly grasp a scientific answer to the ultimate question.

No one doubts quantum theory's accuracy and usefulness. In fact, no theory has ever made more accurate predictions or yielded more world-altering technologies. Nevertheless, scientists have been unable to offer an intelligible description of what the theory actually tells us about the world. The problem is not simply that non-experts find the theory puzzling. The most capable physicists have acknowledged quantum theory's incomprehensibility. Steven Weinberg—who won the Nobel Prize in Physics for his work in quantum theory—admits "to some discomfort in working all my life in a theoretical framework that no one fully understands."[51] In a similar

50 That an exclusively scientific explanation seems ultimately to posit physical existence as a brute fact is not necessarily disqualifying. As we will see, some advocates of other prominent approaches to the puzzle of existence ultimately rest their accounts on an unexplained starting point. Some theists, for example, claim that God created the world, but that there can be no explanation for why God exists in the first place. See Chapter Six.

51 Weinberg, *Dreams of a Final Theory*, 85.

spirit, Carroll laments that "to the chagrin of physicists everywhere, we don't fully understand what the theory actually *is*."[52]

Due to what physicists sometimes call "quantum weirdness," it is difficult to even know what it means for something to exist as a physical thing. We take it for granted that physical things have a specific location. Shockingly, this assumption does not apply at the quantum scale. Except when researchers observe (or "measure") a particle by causing it to interact with larger objects, its properties—like location and momentum—can only be modeled probabilistically. In other words, there is a specific likelihood associated with each state that the particle might be found in if measured at a given time. However, until the particle is actually measured, it is in what physicists call a "superposition," described by a statistical distribution of possible states known as the "wave function."[53] Upon measurement, a particle definitively takes on a unique state (known as "collapse of the wave function").[54]

The idea that physical entities only have definite characteristics when measured is strange enough. But things get even weirder when physicists pose questions about the indefinite state of being known as a superposition. For instance, when measurement causes a collapse of the wave function, why does the particle take on the specific state that it does (rather than any of the many other possible states)? Most physicists answer that there is no reason. The process is genuinely random. Reality at the quantum level is fuzzy or smeared.[55] We do not notice this indeterminacy because the probabilistic nature of quantum events is smoothed out at the level of our experience. This is why a theory based in random events can make such accurate predictions: outcomes are restricted by probabilistic regularities that exhibit great reliability over large enough sample sizes.[56]

Scientists recognize that quantum weirdness challenges our ability to make sense of the world. Nature seems to have one set of rules for apricots, asteroids, and other large ("macro") objects. These are the familiar rules that we intuitively understand. But nature has a different rulebook for electrons, protons, and the other quantum ("micro") objects. The radical difference between the macro and micro worlds would be baffling enough if there were no interaction between them. In fact, however, the two arenas could hardly be more interrelated. After all, macro objects are composed of micro objects. Everything in the macro world—you, me, the stars, and every

52 Carroll, *The Big Picture*, 35.

53 Carroll, *The Big Picture*, 163–64.

54 Paul Davies and J.R. Brown, eds., *The Ghost in the Atom: A Discussion of the Mysteries of Quantum Mechanics* (Cambridge University Press, 1986), 22.

55 Davies and Brown, *The Ghost in the Atom*, 21.

56 Greene, *The Fabric of the Cosmos*, 92–93.

grain of sand—is built from the atomic and subatomic entities of the micro world. Where, then, is the line between them? At what point must we retire one rulebook and pull the other off the shelf? And what exactly provokes the so-called collapse of the wave function?

Although researchers agree on how to *apply* quantum theory in practice, quantum weirdness has generated tremendous controversy over how to *interpret* it. That is, physicists do not agree on what the theory tells us about the fundamental nature of physical things. Three main views dominate the conversation. The understanding of quantum theory described above—known as the "Copenhagen interpretation"—is the most prominent. (It is named after two physicists who pioneered quantum theory in the 1920s at the University of Copenhagen: Niels Bohr and Werner Heisenberg.) Advocates of the Copenhagen interpretation generally accept that quantum events are truly random. In this view, there really is no explanation for why a particle takes on one of the possible states rather than another upon measurement.[57]

However, not everyone agrees. In recent years, some physicists have endorsed competing interpretations of quantum theory. The development of alternative interpretations has been motivated in part by an unwillingness to accept that there is no explanation for why particles have particular properties at a given time. Indeed, Einstein—an early architect of quantum theory—rejected the Copenhagen interpretation for this reason, stating that God does not play dice with the universe.[58]

Competing interpretations of the Copenhagen interpretation have themselves met resistance from other physicists in part because *their* theories also include elements that seem extremely unexpected or bizarre. One alternative proposal—known as the hidden-variable theory—holds that as-yet undiscovered factors (or variables) determine the properties that particles have at particular times. While the hidden-variable theory avoids the randomness and indeterminacy of the Copenhagen interpretation, most physicists reject it due to its own strange implications. In particular, the theory suggests that two particles can instantaneously influence each other no matter how far apart they are, which violates the widely accepted principle that two objects cannot affect each other more quickly than light could travel between them.[59]

The most well-known alternative theory—the many worlds interpretation (or MWI)—is an even more radical departure from the Copenhagen

57 Davies, *The Goldilocks Enigma*, 229.

58 Davies, *The Goldilocks Enigma*, 64; Jim Baggott, "What Einstein Meant by 'God Does Not Play Dice,'" *Aeon*, November 21, 2018, https://aeon.co/ideas/what-einstein-meant-by-god-does-not-play-dice.

59 Travis Norsen, *Foundations of Quantum Mechanics: An Exploration of the Physical Meaning of Quantum Theory* (Springer, 2017), 204–05.

interpretation. The MWI holds that when physicists measure a particle, the universe branches into all of the states described by the wave function. As one MWI proponent puts it, "every time the Universe is faced with a choice at the quantum level, the entire Universe splits into as many copies of itself as it takes to carry out every possible option."[60] In the reality painted by the MWI, the world in which you and I currently live is only one of many worlds that can never interact with one another.

If every possible state described by the wave function becomes real, then why does it *appear* as if only one does? According to the MWI, the illusion that the wave function is probabilistic results from the fact that each version of a person resides in only one world. An individual can only observe one outcome because each of the other outcomes occurs in a different world. Although we are not perceptually aware of it, "the Universe evolves by successively splitting into an ever-increasing collection of different worlds in which everything that can logically [and physically] occur eventually will."[61] Like the hidden-variable theory, the MWI avoids the Copenhagen interpretation's most counterintuitive aspects only at the cost of introducing its own strange elements.[62] Many physicists are simply unwilling to accept the MWI's staggering number of parallel worlds. Our inquiry raises an overarching question about the role that common sense should play in our assessment of theories. It might seem to count against a theory if it violently contradicts deeply held assumptions, but should we assume that reality matches our intuitions?

The critical point for the present discussion is that interpretive debates over quantum theory point to the possibility of certain limits on our scientific understanding. The major interpretations of quantum theory all have similar practical implications in terms of their experimental predictions. However, science is more than a merely pragmatic enterprise. In the end, we want not only to manipulate the world for useful ends, but also to *understand* it. Quantum weirdness raises the possibility that our theories do not allow us to grasp what they mean in terms of what the world is really like. Thus, a scientific approach to the puzzle of existence may suggest that our ability to comprehend the world around us is severely limited.

We have considered a scientific approach's implications for the subject of the most fundamental truths and our ability to grasp reality's genuine nature. Explanations of the world's existence also have implications for humanity's place within the larger picture. For instance, is the notion of purpose built into the cosmos, and, if so, do human beings have a part in its purpose?

60 John Gribbin, *Schrodinger's Kittens and the Search for Reality* (Little, Brown & Co., 1985), 160.

61 John D. Barrow, *The Universe That Discovered Itself* (Oxford University Press, 2000), 406.

62 Barrow, *The Universe That Discovered Itself*, 259–60.

While ancient thought—tracing most famously to Aristotle—envisioned different things in the world as being inclined toward their own distinctive ends, modern science has no place for ideas of that kind. Today's physics centers on the assumption that all things in the world are composed of common basic constituents, answerable to the same behavioral regularities. While ducks behave differently from comets, there is no essential difference between the various things that we encounter. That same leveling applies to human beings, which are not special from a cosmic perspective. From a purely scientific standpoint, we do not live in a universe with intrinsic aims or aspirations. The world may be the stage for stories that matter a great deal to human beings, but they mean nothing to the cosmos, which is as indifferent to us as it is to an erupting volcano or dying star.

Questions for reflection and discussion

1. In your view, what most sharply distinguishes science from other methods of inquiry?
2. How reasonable do you find the *ex nihilo* principle: the idea that nothing comes from nothing?
3. Does the idea of virtual particles help to address difficulties posed by the *ex nihilo* principle?
4. Are problems in the dominant physics theories today relevant in any way to the puzzle of existence?
5. Are scientific theories circular in a way that keeps them from solving the puzzle of existence?
6. What is the best reason to adopt science as an approach to explaining the world's existence?
7. What is the greatest challenge for a scientific approach to the puzzle of existence?
8. What does "quantum weirdness" suggest regarding our ability to understand the world?
9. What does a scientific approach to explaining existence imply about our place in the cosmos?

4.

The Mathematical Universe

The Remarkable Nature of Mathematics

Mathematics is both mundane and extraordinary. We use it every day when setting the alarm clock, paying the bills, and counting eggs to put in our omelet. The everyday ubiquity of mathematics can make it easy to overlook its exceptional features. Consider the nature of our knowledge about things that mathematicians study, such as numbers, sets, and geometrical figures (referred to broadly as "mathematical objects"). So many things in our lives are uncertain, and we yearn for ground that is unshakeable. Mathematics seems to promise just that: claims that are incontrovertibly true. What would it take for you to give up the conviction that two plus two equals four?

Another exceptional feature of mathematics is the apparently necessary nature of its truths. If I look in an office and see a blue chair, I learn that the chair is there at this moment. Yet I also know that this did not have to be the case. If someone had moved the chair out of the room before I arrived, it would not be there, and the chair need not remain in the room forever. Truths about mathematics do not seem contingent in the same way; they could not have been otherwise. Thus, it is often assumed that the angles of a triangle add up to 180 degrees by necessity, in the sense that it did not just happen to be the case. It could not have been any other way. Closely related is the common assumption that the truths of mathematics cannot be altered, and that they have always been the case; that is, that they are both eternal and unchanging. The fact that the angles of a triangle add up to 180 degrees is true today, it was true a million years ago, and it will always be true.

The manner in which we acquire knowledge about mathematics also seems to exhibit remarkable features. We are accustomed to obtaining knowledge about things in the world through the use of our bodily senses. We learn about the specific features of things around us throughout the day by seeing, hearing, touching, tasting, and smelling them. Knowledge about mathematics appears to work in a fundamentally different manner. We do not worry that something we observe in our surroundings tomorrow will reveal that 17 is not really a prime number after all.

The contrast between science and mathematics with respect to the acquisition of knowledge is striking. As discussed in the previous chapter, science investigates claims through observational evidence that can be shared with and confirmed by others. Those proposing scientific theories characteristically offer predictions about the empirical observations that should be expected if they are right. Moreover, any theory that is currently considered to be correct may be called into doubt by subsequent observations.[1] To be sure, the methods of science and mathematics are not wholly dissimilar. For example, mathematicians often show an experimental spirit in making conjectures that look like they might be true before investigating to see if they actually hold up to scrutiny. Nevertheless, unlike scientists, mathematicians can develop proofs based on reasoned arguments that demonstrate through logical steps why a particular proposition must be the case. Once a mathematical result has been proven, other mathematicians may be convinced by the reasoning without reproducing the proof themselves or running repeated experiments to further test it. By contrast, scientific claims are never insulated from the possibility of being discredited by further testing that yields different outcomes.

The distinctive manner in which mathematicians conduct their investigations has led many to believe that numbers and other mathematical objects exist in a real sense that does not depend either on the existence of physical things or on our own thought processes. Indeed, mathematicians often describe the strong sense they have of *discovering* things that are already there. While mathematicians do not trip over the number seven in the same way that a geologist might encounter a new kind of rock in their fieldwork, many nevertheless share the belief that the things they study are "out there" regardless of whether any smart researcher ever points them out.

We have been careful to say that mathematics *seems* to have certain remarkable features. The reason for caution is that philosophers,

1 This does not mean, of course, that the practice of science consists of nothing but running experiments and recording measurements. For instance, scientists draw on various kinds of reasoning in formulating the hypotheses to be tested, and they apply a variety of criteria—such as a preference for simple or elegant theories—in choosing between different theoretical models that imply the same experimental results.

mathematicians, and others continue to debate the true nature of the things that mathematicians study and the ways that we learn about them. We examine this controversy in the section below, "What Are Mathematical Objects?" As discussed further there, many of the views about the nature of mathematics noted above—such as mathematical objects having an existence that is independent of physical things or our ideas about them—are associated with a position known as "Platonism."[2] Not everyone defends Platonism, with some thinkers arguing that mathematical objects are really based either in our sensory observations or in creations of our own minds.

The aim here in noting the apparently remarkable nature of mathematics is not to endorse one particular position in the debate. Rather, the salient point is that many people consider it plausible that mathematics has the remarkable features noted, and that possibility makes mathematics intriguing as a frame for explaining the world's existence.

Establishing that the world exists by necessity—that there is no other possibility—represents a promising approach to the puzzle of existence. Offering a narrative of how the world came to be—as, for example, through the Big Bang or a divine act of creation—is only one way to explain why the world exists. Another intriguing strategy is to show why something *must exist* due to its very nature, and that its existence encompasses—or is equivalent to—the world's existence.

The previous chapter noted two potential limitations on science's ability to explain why there is a physical world. First, scientific theories tend to take physical existence for granted from the start, suggesting that a scientific answer to the puzzle of existence is circular (since it assumes the very proposition that it is supposed to explain). Second, while scientific theories are highly effective in discovering and describing general patterns of behavior, they do not even claim to explain why physical objects follow those patterns rather than others. Mathematics arguably differs in both respects. If, as some believe, mathematical explanations do not depend on the existence of physical things at all, then they do not require a starting assumption of physical existence. In addition, unlike science, mathematics offers explanations that aim to show why certain propositions and relationships could not be other than they are. Necessary truths that are independent of anything physical suggests exciting possibilities for a non-circular solution to explaining why the world exists.

---

2 While Platonists also maintain that the truth of mathematical propositions does not hinge on the sorts of circumstances that we can observe with our senses, they allow that we may need to have certain kinds of experiences to effectively learn principles of mathematics. Nevertheless, the critical Platonist claim is that mathematical objects—and the things that are true about them—have a reality that is independent of physical things or ideas in our minds.

Of course, even assuming that mathematics has the remarkable features discussed, that would not in itself establish that it can solve the puzzle of existence. After all, we do not just want to identify something that exists *in addition to* the physical world. The aspiration is to identify something that is *responsible for* the world's very existence. We could imagine something that has no connection with the physical world whatsoever, and that, therefore, could not possibly be the explanation for its existence. Mathematics can only explain why there is a world if there is some interrelation or interaction between them. Do we have any justification for supposing that there is a deep connection between mathematics and the physical world?

The physicist and mathematician Eugene Wigner suggested a good reason to believe that there is such a connection in his 1960 article "The Unreasonable Effectiveness of Mathematics in the Natural Sciences."[3] The piece shined a light on something familiar to anyone who has taken a science class: numbers are extremely useful in describing the behavior of physical things. In other words, you cannot do science without math.

Recognizing a connection between mathematics and science is hardly new. The Scientific Revolution's key figures demonstrated time and again the power of quantitative expressions as tools for depicting the natural world.[4] Indeed, Galileo authored modern science's motto when he proclaimed that the universe "is written in the language of mathematics."[5] The mathematical science that blossomed during the sixteenth and seventeenth centuries transformed what scientific theories were expected to look like. Before that time, scientific explanations tended to furnish some kind of narrative purporting to portray the means through which one event brought about another. Since that seminal period, however, the scientific community has welcomed theories in the form of mathematical formulas predicting how physical things will behave under specified circumstances. The success of a numerical model in describing outcomes may be accepted as an adequate explanation.[6]

While the usefulness of mathematics in scientific theories provides a reason for suspecting a strong interconnection between numbers and nature, it is even more striking that some mathematical frameworks developed for reasons unrelated to science have turned out to be invaluable for solving

3 Eugene Wigner, "The Unreasonable Effectiveness of Mathematics in the Natural Sciences," *Communications in Pure and Applied Mathematics* 13, no. 1 (1960): 1–14.

4 Barrow, *The Universe That Discovered Itself*, 66, 94.

5 Galileo Galilei, "The Assayer," trans. Stillman Drake, in *Discoveries and Opinions of Galileo* (Doubleday Anchor Books, 1957), 237–38.

6 Michael Heller, *Ultimate Explanations of the Universe* (Springer, 2009), 2; Paul Davies, *The Mind of God: The Scientific Basis for a Rational World* (Simon & Schuster, 1992), 78–79.

problems in physics.[7] A notable example involves Einstein's general theory of relativity. In developing the theory, Einstein incorporated work that had been done on the geometry of multidimensional curved spaces by the mathematician Bernhard Riemann. Although Riemann—who died 13 years before Einstein was born—was not motivated by anything to do with general relativity, his mathematical innovations proved very helpful to Einstein's revolutionary work on gravitation and spacetime.[8] Similarly, in developing the Heisenberg uncertainty principle, physicists drew on applied systems of matrices that had been worked out much earlier by mathematicians for unrelated reasons. Examples like this suggest that mathematics has an integrity of its own while also being interrelated with the world around us.

The role of mathematics in science is so familiar that it is easy to take for granted. Upon reflection, however, the indispensability of mathematical formulas in physics and other subfields in science may seem quite surprising. Given significant differences in the way that scientists and mathematicians conduct their research, it is far from obvious why the two fields should turn out to be so interconnected in practice. For this reason, many scholars feel that the usefulness of mathematics in science presents a mystery.

The essential role that mathematics plays in science raises questions about the nature of the objects that the two fields study. Commentators have taken a variety of perspectives on the subject. One response—common among practicing scientists—is to skip over questions about science's relation with mathematics and to just get on with doing the research. Wigner himself adopted this approach. Having called attention to the curious relation between mathematics and science, he declined to venture an explanation for it. Among those more inclined to speculate on the underlying character of physical and mathematical objects, one prominent approach assigns primary importance to physical existence, with mathematics entering the picture only in a derivative manner. From this standpoint, mathematics is merely a skill or language that we develop to suit our own purposes, such as more efficiently navigating the world around us. Mathematical objects do not have an independent existence of their own. A competing position holds that mathematical and physical objects exist independently of each other. Both kinds of objects stand on their own foundation, but they occupy separate realms.[9]

7 Davies, *The Mind of God*, 151.

8 Penelope Maddy, "How Applied Mathematics Became Pure," *The Review of Symbolic Logic* 1, no. 1 (2008): 20; Weinberg, *Dreams of a Final Theory*, 153–54.

9 Indeed, those supporting this view may deny or minimize the import of Wigner's claim that mathematics is extraordinarily useful in science. They note, for example, that there are arenas—such as weather prediction—where mathematics is not as powerful in supporting accurate predictions as is often assumed to be the case in other scientific fields.

The relation between physical and mathematical objects has implications for the puzzle of existence. If mathematics is subordinate to physical things—if it comes into the picture in only a secondary or derivative way—then it cannot possibly explain why physical things exist in the first place. At the same time, even if mathematical objects exist independently of physical things, they still cannot explain the world's existence if mathematical and physical objects are so distinct that they are like ships passing in the night. Mathematics can only explain the world's existence if (1) mathematical objects exist independently of physical things; (2) mathematical principles are necessarily true; and (3) mathematics and the world are interconnected. In the next section, we examine what a view of that kind might look like.

Mathematical Answers to the Puzzle of Existence

The idea that mathematics is the bedrock of the cosmic order is almost as old as philosophy itself. Pythagoras (c. 570–490 BCE)—who espoused that idea—was born only about a half-century after the first known Western philosopher, Thales. Born in Samos, Greece, off the coast of Asia Minor, Pythagoras reportedly traveled as a young man to Miletus, where he met Thales (and possibly another of the great Milesian philosophers, Anaximander). After traveling to Babylon and other places, he established a school in Croton (in present-day southern Italy).[10] While we do not know what Pythagoras learned from his various hosts, it is tempting to suppose that the Milesians intensified his thirst to identify the *arche* (or fundamental basis of reality),[11] while the Babylonians—who understood mathematics remarkably well—turned his sights toward numbers as reality's very foundation.

As is often the case with respect to ancient thinkers, little is known with any assurance about Pythagoras' life. The uncertainty is exacerbated both by Pythagoras' refusal to write anything and by his school's insistence on secrecy regarding the master's teachings.[12] Historians continue to debate details about Pythagoras, including whether he was an actual person or only a legendary figure. There is not even any evidence showing conclusively that he proved the geometrical theorem famously tied to his name.[13] Nevertheless, despite the dearth of evidence regarding the histori-

10 Morris Kline, *Mathematical Thought from Ancient to Modern Times* (Oxford University Press, 1992), 28–29.

11 The Milesian quest to identify the *arche* is discussed in Chapter Two.

12 Hussey, *The Presocratics*, 61.

13 Peter Adamson, *Classical Philosophy: A History of Philosophy without Any Gaps* (Oxford University Press, 2014), 23–24.

cal Pythagoras, the ideas attributed to him and his followers are significant due to their historical impact. Regardless of what Pythagoras himself may have believed, people calling themselves Pythagoreans exerted influence in ancient Greece.[14] Indeed, Aristotle wrote a book (now lost) about Pythagoras and some of Plato's students considered themselves Pythagoreans.[15]

It is not the Pythagoreans' competence in mathematics alone that makes them of such great interest. The Babylonians had developed sophisticated mathematical ideas much earlier, applying them across a range of activities from astronomy to land measurement.[16] Well before Pythagoras, mathematics had already been studied by other Greeks, who saw it as a distinctly abstract body of knowledge.[17] Rather, what makes the Pythagoreans salient to our exploration is that mathematics meant more to them than a set of useful tools or intriguing intellectual challenges. For the Pythagoreans, numbers represented the constituents of the cosmos itself. They were the irreducible reality that remained when the inessential was eliminated.[18]

Fundamental for the Pythagoreans was their conviction that the cosmos manifested a deep harmony rooted in the patterns of mathematics. This belief resonates with Wigner's observation about the extraordinary usefulness of mathematics in the natural sciences. To be sure, the Pythagoreans preceded the development of predictive scientific models employing complicated mathematical formulas. That leap would come much later with the Scientific Revolution. Nevertheless, the basic insight animating Pythagorean thought is similar to one that drives modern physics: the cosmos is ordered according to numerical patterns. No less than today's scientists, the Pythagoreans reveled in discovering observable connections between the world and mathematics. Most famous in this respect was their legendary discovery of the role that quantitative ratios play in music. Producing harmonies, the Pythagoreans observed, requires a ratio of whole numbers. For example, musical notes an octave apart exhibit a two-to-one ratio (as in the length of the strings vibrated).[19]

It is easier to overlook the Pythagoreans' significance because they bundled their views about mathematics together with teachings that people today think of as falling in different domains. They professed belief in an immortal soul that ascended to higher states in subsequent lives,[20]

---

14 Kline, *Mathematical Thought from Ancient to Modern Times*, 28.

15 Adamson, *Classical Philosophy*, 25.

16 Hussey, *The Presocratics*, 65–66.

17 Kline, *Mathematical Thought from Ancient to Modern Times*, 29.

18 John D. Barrow, *Pi in the Sky: Counting, Thinking, and Being* (Clarendon Press, 1992), 251–52.

19 Charles H. Kahn, "Pythagorean Philosophy before Plato," in *The Pre-Socratics: A Collection of Critical Essays*, ed. Alexander P.D. Mourelatos (Princeton University Press, 1993), 177.

20 Adamson, *Classical Thought*, 29.

for instance, and prescribed ritualistic practices on everything from permissible foods to a prohibition on looking into a mirror that is next to a lamp.[21] More closely related to their understanding of mathematics itself, the Pythagoreans also attached symbolic meaning to particular numbers. Perfect numbers—those that are the sum of their divisors, like the number 6, which equals 1 plus 2 plus 3—were accorded special status. They viewed the number 10 as sacred because it was the sum of the first 4 whole numbers. More generally, the Pythagoreans assigned special significance to natural numbers (the positive integers: 1, 2, 3, and so on), believing that the relations defining geometrical figures were all expressible as the ratios of such numbers.[22]

The Pythagoreans' astronomical investigations illustrate how their belief in mathematical harmony generated conclusions about the cosmos. The Pythagoreans assumed that there were ten perfect heavenly spheres based on that number's sacred status.[23] Yet they only managed to observe eight heavenly spheres: five planets in addition to the Earth, Moon, and Sun. Due to the significance that they accorded to the center of a circle, the Pythagoreans presumed the existence of a ninth heavenly sphere that was an invisible fire around which the other bodies revolved. Since this central fire plus the eight observable bodies left them one shy of the requisite number, the Pythagoreans posited a tenth heavenly body, which they called the Counter-Earth.[24]

Many of the Pythagoreans' beliefs would now be viewed as implausible or outside the scope of mathematics altogether. Nevertheless, their core belief that mathematics played a vital role in understanding the natural world remains resonant. Although researchers today do not base conclusions on sacred numbers, the Pythagorean impulse to understand the physical world through mathematical relationships foreshadows Galileo's assertion that the universe is written in the language of mathematics.

The Pythagorean spirit remains alive and well in the work of the physicist and cosmologist Max Tegmark (b. 1967). Tegmark's "Mathematical Universe Hypothesis" (MUH) is the most well-known contemporary mathematical theory of existence. Examining the MUH will further illuminate the general appeal of a mathematical approach to explaining the world's existence while showing what such a theory can look like when fleshed out.

21 Hussey, *The Presocratics*, 64–65.

22 W.S. Anglin and J. Lambek, *The Heritage of Thales* (Springer, 1995), 33. Due to this belief, the Pythagoreans faced a crisis upon discovering irrational numbers, which are expressed as decimals that do not repeat or terminate. Kline, *Mathematical Thought from Ancient to Modern Times*, 32–33.

23 Davies, *The Mind of God*, 94–95.

24 Charles H. Kahn, *Pythagoras and the Pythagoreans: A Brief History* (Hackett, 2001), 25–26.

Let us begin with the root motivation underlying the MUH, which explicitly invokes Pythagoras in proclaiming that the cosmos exhibits mathematical harmony and order.[25] While the central idea that Tegmark draws on has ancient roots, modern science has revealed manifestations of the interconnection between mathematics and the cosmos that ancient thinkers could not have imagined, from electromagnetism to radioactivity and biochemical reactions.[26] As Tegmark notes, the universe's basic constituents can be expressed as "pure numbers" without reference to the units of measurement that we assign to them (such as inches or kilograms). For example, a proton's mass is 1836.15267 times greater than that of an electron.[27] Moreover, electric and magnetic fields can be described by three numbers, referring to strength and direction, and, when unified into an electromagnetic field, they can be represented by six numbers at each point.[28] In Tegmark's words, even space, "the very fabric of our physical world ... is a purely mathematical object in the sense that its only intrinsic properties are mathematical properties—numbers such as dimensionality, curvature and topology."[29] In effect, there are "a bunch of numbers at each point in spacetime," and this idea, Tegmark maintains, is "telling us something not merely about our description of reality, but about reality itself."[30]

Another good reason to embrace a mathematical answer to the puzzle of existence, Tegmark maintains, is that it provides the most compelling response to a question that we discussed earlier: Why is mathematics useful in the practice of science? In addressing this mystery, Tegmark radically reconceives the relation between mathematics and science. It is often assumed that mathematics and science are essentially different disciplines, with the first studying the abstract relationships among mathematical objects and the other describing the physical world. The challenge, then, is to explain why two such seemingly distinct fields are inextricably linked in practice. Tegmark, however, denies that mathematics and science are separate in the first place. Instead, he contends, mathematics proves useful in describing reality because reality *is* mathematics. Correct scientific theories move us closer to an understanding of the actual mathematics that is the universe. As Tegmark writes, "our successful theories are not mathematics approximating physics, but mathematics approximating mathematics."[31]

25 Max Tegmark (individualized entry), in Max Tegmark, Piet Hut, and Mark Alford, *Foundations of Physics* 36, no. 6 (2006): 769.

26 Max Tegmark, *Our Mathematical Universe: My Quest for the Ultimate Nature of Reality* (Knopf, 2014), 249.

27 Tegmark, *Our Mathematical Universe*, 250–51.

28 Tegmark, *Our Mathematical Universe*, 279.

29 Tegmark, *Our Mathematical Universe*, 253. "Topology" concerns space's overall shape and structure.

30 Tegmark, *Our Mathematical Universe*, 278.

31 Max Tegmark, "The Mathematical Universe," *Foundations of Physics* 38 (2008): 107.

Tegmark bolsters his case for a fundamentally mathematical worldview with an argument based in the idea of objective truth. To begin, Tegmark posits what he considers a relatively uncontroversial view, the "External Reality Hypothesis" (ERH), which holds that the world's existence does not depend on human beings. The world would persist even if every person vanished.[32] Tegmark unfolds the argument by working through the ERH's implications. First, to say that the world exists independently of humanity means that it can be described objectively. An objective description of the world includes only that which is indispensable to the way that anyone would describe reality, regardless of their particular culture. Thus, any part of a description that depends on a particular observer's vantage point must be discarded. When everything inessential is stripped away, we are left with an objective description, one that could be processed by a computer or understood by intelligent creatures from other planets.[33]

One might guess that Tegmark identifies an objective description of reality with scientific theories, but that is not the case. Even the best theories in the natural sciences, he notes, contain unnecessary descriptions. Along with their quantitative formulas, theories in physics typically include verbiage explaining "how the equations are connected to what we observe and intuitively understand."[34] As Tegmark writes: "When we derive the consequences of a theory, we introduce new concepts—protons, molecules, stars—because they are convenient."[35] Since these concepts are culturally bound, they must be jettisoned as excess baggage. After all, societies adopt a variety of terms and concepts to characterize the world. Communities give names to things based on what they find notable or useful, but other cultures (human or otherwise) would make different choices.[36] Thus, the language that scientists use to embellish their theories can be eliminated without fundamentally altering our descriptions of the external reality itself.[37]

What, then, is essential to an objective description of the universe? For some time, scientists have aspired to discover a single, unified quantitative model that predicts any kind of physical behavior. Tegmark shares the dream of discovering such a "theory of everything." For Tegmark, an objective description is the quantitative expression that constitutes the theory of

32 While Tegmark may be right that many people accept the External Reality Hypothesis, there are a number of grounds on which it might be challenged. For example, some would argue that the world is nothing more than a projection of minds—whether human or divine—or that there is no way to establish the external world's existence even if one exists.

33 Tegmark, "The Mathematical Universe," 102.

34 Max Tegmark, "Shut Up and Calculate" (2007), 1, accessed December 8, 2024, https://arxiv.org/abs/0709.4024.

35 Tegmark, "Shut Up and Calculate," 1.

36 Tegmark, "Our Mathematical Universe," 256.

37 Tegmark, "Shut Up and Calculate," 2.

everything. Based purely on that quantitative expression, a supercomputer, or an "infinitely intelligent mathematician,"[38] could "calculate how the state of the universe evolves over time without interpreting what is happening in human terms."[39]

Tegmark further argues for the MUH on the grounds that it provides a solution to a problem that has long preoccupied philosophers: how to explain the properties that things have without falling into an infinite regress. Properties are the various attributes or qualities that objects exemplify. To describe an item that you bought at the store, you would talk about its properties. Perhaps the shirt you bought was orange and made of cotton, and the suitcase was lightweight and water resistant. One of the most fundamental questions about the world concerns the reason for the properties that things have: What ultimately explains why different things have the particular properties that they do?

Tegmark thinks the answer has been elusive because people give explanations that point to additional questions. The process by which explanations engender additional questions is known as a "regress," and an "infinite regress" occurs when this process has no end. Some thinkers, including Tegmark, believe that an infinite regress is not a satisfying explanation because it never reaches a foundation or stopping point. Tegmark refers to this as the "classic problem of infinite regress, where every explanation of a statement in human language must be in the form of another unexplained statement."[40] However, Tegmark assumes that there is an explanation for the properties that objects have, which means that the chain of explanations must have a terminus. Science, as conventionally understood, cannot stop the regress, Tegmark says, because its explanations for properties simply refer back to the properties of other things, thus generating the forbidden infinite regress. By contrast, mathematics can provide a satisfying explanation for the properties that things have because, unlike science, it is rooted in necessity. Since mathematics can show why its truths could not be other than they are, its explanations do not generate an infinite regress the way that conventional science does. Thus, the necessary truths of mathematics are responsible for the properties of all things, from rectangles and parabolas to umbrellas and platypuses.[41]

Now, anyone proposing that mathematics explains the world's existence must address a critical question: What is the relation between mathematics

38 Tegmark, "Our Mathematical Universe," 259.

39 Tegmark, "Shut Up and Calculate," 1.

40 Tegmark, "On Math, Matter, and Mind," 769.

41 Given Tegmark's notion of equivalence, there is ultimately no difference between mathematical and physical properties.

and the physical world? One can imagine different answers to that question. For instance, one might say that mathematics has a fundamentally distinct character from the world, but nevertheless is responsible for generating its existence. In a broad sense, theistic frameworks use a similar approach, claiming that something fundamentally distinct from the world—a divine being—created the world from nothing.[42] Tegmark describes a different kind of relation between mathematics and the world. In his theory, mathematics does not create or generate the world. Instead, Tegmark draws an equivalence between mathematics and the world, collapsing the distinction between them. The cosmos is not some separate or distinct thing that grows out of mathematics: it *is* mathematics.[43]

Because the MUH eliminates the distinction between mathematics and the world, how the theory views the essential character of mathematics has far-reaching implications. Some theorists believe that mathematics concerns the properties of *individual* numbers, geometrical figures, or other mathematical objects. But Tegmark denies that mathematics studies the nature of mathematical objects taken in isolation, insisting that its real subject is the *relations between them*.[44] In this view, what matters are mathematical structures, which are composed of multiple mathematical objects.[45]

For Tegmark, a mathematical structure is described by a quantitative expression within a well-defined formal system, which includes axioms and rules that can be employed to produce true statements. Axioms are first principles that are not proven, but assumed to be true. The rules of a formal system dictate how one may apply the axioms to prove other propositions. A formal system's axioms and rules, used properly, give rise to theorems: mathematical claims that inexorably follow from the established starting points.[46] Tegmark stresses the deep connection between formal mathematical systems and computer programs, noting that both operate according to a prescribed set of rules which determine the relationships that necessarily follow from the rules. Thus, computer programs can generate the mathematical structures of particular formal systems, including geometrical figures (like cubes, spheres, and cones), numerical structures (such as the natural numbers, the integers, and the rational numbers), networks, and topological spaces.[47]

42 Theistic explanations for the world's existence are the subject of Chapter Six.

43 Tegmark, *Our Mathematical Universe*, 260.

44 Tegmark, *Our Mathematical Universe*, 259.

45 Colin Hamlin, "Towards a Theory of Universes: Structure Theory and the Mathematical Universe Hypothesis," *Synthese* 194 (2017): 571.

46 Max Tegmark, "Is 'The Theory of Everything' Merely the Ultimate Ensemble Theory?," *Annals of Physics* 270, no. 1 (1998): 8.

47 Tegmark, *Our Mathematical Universe*, 268.

Given the assumption that reality is mathematics, and that mathematics concerns structures, Tegmark reasons that our entire universe is itself a complex mathematical structure. Elaborating on the meaning of equivalence, Tegmark proposes that two things are equivalent if they can both be fully described by the same mathematical expression. Our universe's theory of everything is a mathematical expression that describes a mathematical structure while also describing the universe in which we find ourselves. Thus, our universe is equivalent to a mathematical structure, namely the one specified by the theory of everything. Since Tegmark considers the theory of everything an objective and complete description of our universe, it follows that everything about the universe traces back to the theory of everything, including, for instance, the elementary particles, the way gravity works, and even human psychology.

This same set of premises leads Tegmark to another radical claim: that our universe exists within a multiverse containing an extremely large number of universes that differ from one another. If everything is mathematical, and everything about mathematics could not possibly be any other way, then there is no gap between possibility and reality. The MUH permits no unrealized alternatives. Anything that qualifies as a mathematical structure really exists. Since there are countless mathematical structures with different properties, there must be countless universes with varying properties. As Tegmark puts it, the multiverse comprises "a diverse zoo of mathematical structures."[48] We can bring to mind some idea of what other universes might be like by imagining differences from our own. Some universes are like ours with only "small modifications,"[49] while others are unlike anything we know. They might have no light or gravity, a different number of spatiotemporal dimensions or, perhaps, no spacetime at all.[50]

Appealing to a preference for simpler theories, some object to the MUH on the grounds that Tegmark's multiverse is the farthest thing from simple. Instead of enhancing our understanding, these critics charge, the MUH makes things unnecessarily complicated by introducing a huge number of additional universes that also require explanation. Tegmark recognizes the importance of simplicity, but argues that his opponents misunderstand its meaning. While the critics view simplicity as minimizing scale or quantity, Tegmark views simplicity as concerning the amount of information that is needed to describe a system. On this conception, one thing is simpler than another if it can be specified with a shorter algorithm.[51] Simplicity's enemy

48 Tegmark, *Our Mathematical Universe*, 323.

49 Tegmark, *Our Mathematical Universe*, 323.

50 Tegmark, *Our Mathematical Universe*, 324–25.

51 Tegmark, "The Mathematical Universe," 121.

is not large numbers of things, but arbitrariness regarding which particular things are included. In this respect, an "entire ensemble is often much simpler than one of its members."[52]

To illustrate Tegmark's understanding of simplicity, consider how much easier it is to list the whole numbers from 1 to 100 than to recall the numbers 14, 24, 37, 43, 59, 68, and 91. The first set of numbers is larger, yet easier to recite from memory. The reason is that the first set can be described with a concise instruction: begin with the number one and keep adding 1 to the previous number until you reach 100. No such simple instruction is available for the shorter list of numbers because each item on it must be individually specified.

With the concept properly understood, Tegmark argues that the positing of a multiverse makes his theory remarkably simple. If only one universe existed, then describing it would require a tremendous amount of information. Its arbitrary features would all have to be individually specified. By contrast, the multiverse can be described with relative conciseness, because it contains a universe corresponding to each mathematical structure. Once Tegmark specifies the meaning of a mathematical structure and recognizes each one as an existing universe, the rest necessarily follows and the arbitrariness disappears.

Tegmark's collapsing of the distinction between mathematics and science has radical implications for how we can learn about the world. In the conventional view, science describes the world based on empirical observations, and there is no way that we could discern those patterns apart from measuring the ways that particular things actually interact with each other. For Tegmark, however, the universe is a mathematical structure, which means that its properties follow necessarily from the nature of the particular structure that it is. As a mathematical structure, the universe can be studied via tools associated with that discipline. Once we know our universe's theory of everything, in principle we should be able to do everything that science does today in the same way that we do mathematics: without reliance on empirical observations. The principles of chemistry could be gleaned from physics, those of biology from chemistry, those of psychology from biology, and so on.[53] We could even learn about all of the other universes in the multiverse, since each one is its own mathematical structure.

The idea that we can learn many things about a subject through deductive reasoning, independent of experience, is familiar to us from mathematics. Once we know the axioms and rules of a mathematical system—as in arithmetic or geometry—we can reach myriad conclusions about things

52 Tegmark, "The Mathematical Universe," 121.
53 Tegmark, "The Mathematical Universe," 103.

that are true within that system. Knowing the definition of triangles, we can draw conclusions about the properties they must have without having to conduct laboratory experiments. What makes the MUH revolutionary is that it applies methods familiar in one domain (mathematics) to a domain commonly thought to be essentially different (science), because it assumes that they are really one and the same thing. Tegmark's remarkable claim is that a universe's properties are deducible from its theory of everything, just as a geometrical figure's properties follow from its definition.

Tegmark, a professional physicist, is not denying the *current* need for empirical science. There are at least two limitations on our present knowledge that prevent researchers from replacing science with mathematics. First, we must make empirical observations to identify the theory of everything for the particular universe in which we find ourselves. Second, even if we knew our universe's theory of everything, we could not say exactly how it explains all of the properties that our universe manifests. That is, we do not yet understand how a mathematical structure's properties account for—or are equivalent to—the properties of entities as described by today's physics theories. Thus, for now, the theories developed by empirical science provide us with a "handy user's manual" for how things work in the particular mathematical structure that is our home.[54] Nevertheless, the MUH asserts that the properties of the things around us are in principle derivable from the theory of everything. Figuring out how to do this would be the "ultimate triumph of physics."[55]

While conceding that we are presently a long way from replacing science with mathematics, Tegmark offers tentative ideas about how we might in the future be able to discern the principles that explain how mathematical relations constitute the things around us. In this vein, he suggests that we begin by examining symmetries. A symmetry is a property that remains unchanged upon undergoing a certain kind of transformation. For instance, an unmarked sphere remains unchanged when rotated on an axis, and an object with mirror symmetry looks the same upon being reflected from left to right.[56] Symmetries are promising because they play such an important role in both mathematics and science. As Tegmark notes, symmetries "are among the very few types of properties that every mathematical structure possesses," and they "can manifest themselves as physical symmetries."[57] Contemporary physicists focus heavily on symmetries, which are crucial for

---

54 Tegmark, "The Mathematical Universe," 140.

55 Tegmark, *Our Mathematical Universe*, 237.

56 Tegmark, *Our Mathematical Universe*, 266.

57 Tegmark, *Our Mathematical Universe*, 337.

researchers seeking to unify quantum mechanics and general relativity.[58] Although Tegmark's brief discussion of symmetries does little more than gesture toward what it might look like for mathematics to explain physical properties, it illustrates the seriousness with which his theory takes the notion that mathematics and the world are really one and the same.

What Are Mathematical Objects?

We began this chapter by noting apparently remarkable features of mathematics that make it an intriguing vantage point from which to explain the world's existence. Mathematics seems to offer knowledge about things that are necessarily, certainly, eternally, and inalterably true. In addition, we seem to acquire this knowledge about mathematics in ways that do not depend on making empirical observations. Moreover, many consider it plausible to suppose that mathematical objects do not depend on physical things for their existence. Assumptions of this kind raise exciting possibilities for developing a mathematical approach to explaining the world's existence, and the last section examined how Tegmark does just that.

However, all of the claims that we have noted about the nature of mathematics are controversial. Debate about them spans nearly the entire history of philosophy and continues to preoccupy thinkers today. As noted earlier, the position in this debate that most closely tracks the views about mathematics with which we began the chapter is known as "Platonism."[59] Platonists maintain that mathematical objects have a real existence that does not depend either on things in the physical world or productions of our own minds. Advocates of the position commonly refer to the things studied by mathematicians as "abstract objects," although it is easier to say what these kinds of objects *are not* than to articulate precisely what they are. Our everyday experience is so immersed in the individual items that we perceive with our bodily senses that we strain for words in describing things which do not suggest a physical reality. Consequently, abstract objects are commonly defined in terms that explicitly draw contrasts with physical things. We say that abstract objects do *not* exist in space and time and that they do *not* causally interact with objects in space and time, but that they nevertheless objectively exist.

58 Davies, *The Goldilocks Dilemma*, 159–63.

59 While the position known as Platonism is inspired by Plato and reflects his thought in important respects, it cannot be assumed that Plato himself would have embraced all of the views endorsed by contemporary "Platonists." Likewise, it cannot be assumed that philosophers who identify themselves as "Platonists" in debates over the nature of mathematical objects agree with the entire body of ideas associated with Plato.

Since Platonism depends on the claims that mathematical objects really exist but are neither physical nor mental in character, competing positions can be identified by which of these claims they deny and what they propose as an alternative. One school of thought—which we may refer to as "empiricism"—maintains that mathematical objects have their basis in the physical world. Empiricists naturally do not mean by this that there is a large number fifteen made of iron perched on a mountaintop somewhere, but, rather that we learn about numbers by observing individual things in the world. For instance, we can say that a grape on a vine together with another one on the same vine share the property of two-ness. The grapes are not the same thing as the number, but two-ness is dependent on the existence of particular things that embody the property.[60]

While Platonism draws appeal from its resonance with the ways that we think and talk about mathematics, some find it unable to provide a convincing explanation of how we gain mathematical knowledge. We claim to know many things about mathematics, yet learning about a subject seems to require making some kind of contact with it. What kind of contact could we have with something that has an existence utterly separate from the world in which we live? An important part of what makes empiricism attractive to some observers is that it has a ready explanation for our ability to learn about mathematics. If mathematics has its basis in the physical world, then our facility in acquiring mathematical knowledge may not seem terribly surprising or mysterious, since we can perceive physical things around us with our bodily senses.

Another kind of contact that we can make with the objects of our knowledge is mental in character. That is, we can know things about the contents of our own minds. We learn things through reflection on the ideas that we conceive and discuss. A view—which we may refer to as "conceptualism"—characterizes the things studied by mathematicians as mental representations that are effectively products of our own minds. Versions of conceptualism adopt different characterizations of the mind's activity in generating mathematical objects. For instance, some conceptualists think that mathematical objects originate in social customs, or in felt needs to address various practical problems that arise. Others associate mathematical ideas with faculties built into human psychology that impose certain patterns of thought on the way that we cope with the things around us. Some conceptualists insist that our mathematical ideas do not come from contingent psychological mechanisms that happen to occur in *some* people's minds but, rather, are rooted in what is common and universal to *all* minds

60 Øystein Linnebo, *Philosophy of Mathematics* (Princeton University Press, 2017), 90–91.

capable of engaging in mathematics.[61] Nevertheless, regardless of the details, conceptualists agree that we generate mathematical objects ourselves rather than discovering them as entirely mind-independent things.

In explaining how we acquire mathematical knowledge, both empiricism and conceptualism rely on means of acquiring knowledge—sensory perceptions and ideas in our minds, respectively—that are familiar and relatively uncontroversial. Nevertheless, Platonists argue that neither empiricism nor conceptualism can account for the full range of ways that we think and talk about mathematical objects. Empiricists claim that knowledge of mathematical objects comes from our senses, but mathematicians work with ideas and concepts that do not seem to have analogues in the physical world. For example, it is hard to see how negative numbers can be the product of our encounters with particular, concrete things in the physical world. How can we see or touch *negative* five peaches or water bottles?

Even if empiricism can explain our knowledge of negative numbers—perhaps on the grounds that we can perceive the removal of objects—certain complex mathematical ideas are even more problematic. Consider imaginary numbers, for instance, which are defined as the square root of a negative number. Imaginary numbers play vital roles in algebra, calculus, and many other aspects of mathematics, yet it is difficult to envision how we could encounter imaginary numbers in the physical world. A negative number multiplied by itself is always positive and, thus, there cannot be any particular number whose square yields a negative number. While mathematicians use imaginary numbers in describing certain physical phenomena—such as the swinging of a pendulum—we cannot directly observe them. Thus, we can only understand them, Platonists contend, through reflection on abstract objects. The same applies to various other ideas that are not generally familiar to laypersons but that mathematicians work with, such as a 20-dimensional sphere.

Platonists argue that conceptualism does not accurately reflect the nature of mathematics either. They note that many mathematicians experience their work as proving eternal truths that would be true regardless of our ideas about them.[62] As the mathematician Timothy Gowers writes, "almost all mathematicians who successfully prove theorems feel as though they are making discoveries."[63] Similarly, the astrophysicist Piet Hut describes the process of solving a mathematical problem as stumbling "upon it in a way that is rather similar to the way we stumble over a chair. The resistance

61 Linnebo, *Philosophy of Mathematics*, 78.

62 Linnebo, *Philosophy of Mathematics*, 11.

63 Timothy Gowers, "Is Mathematics Discovered or Invented?," in *Meaning in Mathematics*, ed. John Polkinghorne (Oxford University Press, 2011), 3.

that mathematical objects show to our attempts to prove what later turns out to be false is akin to the resistance that physical objects show when we try to wish them away."[64] Conceptualism's treatment of mathematics as just another branch of human psychology is difficult to square with mathematicians' sense that they are uncovering external truths.

The subset of conceptualists who ground mathematics in universal conditions of rational thought respond that their view can explain why people share the same mathematical concepts. Mathematics, they say, does not involve empirical psychology that is based in observing the contingent and varying behavior of particular individuals. From this conceptualist perspective, it is hardly surprising that mathematicians claim to discover objective truths when they identify previously unknown ideas that are built into the very fabric of our thought. Nevertheless, Platonists reject this version of conceptualism as well, arguing that it is implausible to suppose that mathematical theorems cannot be true unless and until they are proven by human beings.[65]

Another school of thought—known as "fictionalism"—denies that mathematical objects exist at all. Fictionalists analogize mathematics to stories and other narratives that human beings fabricate. We create fiction for various reasons, ranging from entertainment to convenience or other useful ends. According to fictionalists, the number 19 is no more real than dragons or Batman. We recognize the difference between fantasy and reality when we say that something is true only from the vantage point of a fictional narrative. One may say that "Sherlock Holmes lives at 221B Baker Street" without committing to the claim that Sherlock Holmes literally exists. Similarly, one may say that "2+2=4" without committing to the existence of 2 and 4. But a person who believes that Sherlock Holmes is really an English detective is seriously confused. A fairy tale may elicit powerful themes and morals, and mathematics might be a useful tool in our everyday lives, but claims about things in fictional narratives are not *really* true because they are about things that do not actually exist.

Why would anyone endorse the strange-sounding notion that mathematics is a fiction? The motivation for adopting surprising positions is often to avoid bigger problems with the competing positions. After acknowledging that fictionalism "can seem a bit crazy," the philosopher Mark Balaguer observes that "the appeal of fictionalism starts to emerge when we realize what the alternatives are." "The main argument for fictionalism," he notes, "proceeds essentially by trying to eliminate all of the alternatives

64 Piet Hut (individualized entry) in Max Tegmark, Piet Hut, and Mark Alford, "On Math, Matter, and Mind," *Foundations of Physics* 36, no. 6 (2006): 777.

65 Linnebo, *Philosophy of Mathematics*, 78–79.

to fictionalism."[66] In particular, fictionalists refuse to accept the existence of abstract objects—thereby rejecting Platonism—while also denying the plausibility of either empiricism or conceptualism. Having ruled out that mathematical objects exist either as abstract objects or as anything other than abstract objects, fictionalists simply reject the existence of mathematical objects altogether.

Refuting fictionalism, Gottlob Frege (1848–1925)—a Platonist mathematician and philosopher—developed a particularly influential argument defending the existence of mathematical objects. Frege began by noting that people often make claims in the form of a sentence with a subject and a predicate (or phrase describing the subject). In the sentence "The apple is red," the "apple" is the subject and "red" is the predicate, which indicates something about the apple. While there is nothing surprising in this observation, Frege's argument rests on two additional premises. First, he said that a claim expressed in the form of a subject and predicate can only be true if the things referred to by the sentence actually exist. For example, the sentence, "The apple is red," can only be true if there exist both an apple and the property of redness, which make the sentence true. Second, Frege assumed that at least some of the claims people make about mathematics *are* true (as, for instance, when one asserts that the number 17 is prime). Based on these premises, Frege reasoned that numbers had to exist, because otherwise there could not be any true statements about them.[67]

Frege's defense of mathematical objects reflects distinctive challenges confronting Platonism. Since Platonists insist that numbers are neither physical nor mental, they cannot rely on familiar strategies for supporting their views, such as empirical observations or ideas in our minds. Instead, Platonists often argue that accepting their positions is the only way to make sense of certain ways that we think and talk about the world. In addressing questions about existence and truth, they insist, it is not only what we can perceive that matters but also what we can conceive, comprehend, and convey to others. After all, these things are also part of the way that we encounter the world that must be explained. Platonists buttress their claims by showing they are the only way to account for certain ideas that we are able to form and understand.

This mode of reasoning harks back to Plato himself, who used it to show that certain kinds of things—like numbers—are real, even though we cannot perceive them with our bodily senses. The very fact that we can

66 Mark Balaguer, "Fictionalism in the Philosophy of Mathematics," in *The Stanford Encyclopedia of Philosophy*, ed. Edward N. Zalta and Uri Nodelman (2023), Introduction and § 1.1, accessed December 10, 2024, https://plato.stanford.edu/archives/spr2023/entries/fictionalism-mathematics/.

67 Linnebo, *Philosophy of Mathematics*, 10.

understand certain ideas suggests that they have an independent reality. To illustrate, suppose a friend reports that she encountered three dogs on her way to a three-course meal with three friends. Not only would you understand what your friend meant by "three" in relation to dogs, dishes, and people, but you can also understand and talk about the idea of three-ness apart from any particular things to which the number applies. You can grasp what the dogs, dishes, and people all share in manifesting the property of three-ness. The best way to make sense of this, the argument goes, is that three-ness exists independently of dogs, dishes, people, or any other particular physical things.

The same style of reasoning applies to other kinds of mathematical objects, such as geometrical figures. In our everyday lives, we perceive many objects with roughly the shape of a triangle, while also realizing that none of them is an absolutely perfect triangle. Any particular thing that we experience in our surroundings has lines that are not precisely straight. No matter how tiny, there is a smudge here or a defect there. Our eyes will never see an ideal triangle. Yet we nevertheless know a great deal about perfect triangles and other geometrical figures. Indeed, it is telling that we are able to recognize particular things as imperfect geometrical figures in the first place. How do we know that a figure drawn in the sand is an imperfect triangle unless we already have the idea of a perfect triangle? According to Platonists, the best explanation is that we are tapping into something beyond our experience of the physical world.

Now, any position affirming that we know things about mathematics must explain how we acquire that kind of knowledge. That challenge is especially pressing for Platonism, since it denies that we can learn about mathematical objects exclusively through sensory perception or ideas that are the product of our own minds. Some theorists believe that we can only gain knowledge of things by making contact or interacting with them. Based on that assumption, it seems mysterious how we could acquire knowledge about something that is neither physical nor mental in character.

While there is no single, consensus view among Platonists about the best way to account for our mathematical knowledge, many posit a distinctive intuitive faculty that provides us with access to mathematical knowledge. Here, Platonists often draw an analogy with the sensory capacities that we use to gain knowledge about the physical world. This kind of response resonates with Plato's influential idea that there is more to reality than the physical world, and that human beings are capable of comprehending certain non-physical aspects of reality.[68] It also satisfies the assumption that we

68 Mark Balaguer, *Platonism and Anti-Platonism in Mathematics* (Oxford University Press, 1998), 25.

must make some sort of contact with mathematical objects in order to learn about them (in this case, through an intuitive faculty).

A distinctive approach that some Platonists take to explaining mathematical knowledge is to deny that acquiring knowledge about mathematics requires making any sort of contact with mathematical objects. Along these lines, for instance, Balaguer has defended "plenitudinous" or "full-blooded Platonism" (FBP), which rests on the premise that "all the mathematical objects that logically *could* exist actually *do* exist."[69] In this view, we do not need to make contact with mathematical objects to determine their truth, because we can assume that all of the mathematical objects described by any consistent mathematical theory actually exist. Thus, we can acquire knowledge about mathematical objects simply by concluding that a particular theory is consistent.[70] If FBP is correct, then all consistent purely mathematical theories truly describe some collection of abstract mathematical objects. Whatever the details of their particular accounts, Platonists insist on the central claims that mathematical objects really exist—as abstract objects—and that we can obtain genuine knowledge about them.

From Pythagoras and Plato to the present day, philosophers have sought to explain the remarkable nature of mathematics. While not all those engaged in the debate are moved by the challenge of explaining the world's existence, their arguments have ramifications for it. Theorists like Tegmark who identify mathematics as the foundation of reality must defend their views from the claims that mathematical objects depend for their existence on the physical world or our own minds, or that they do not exist at all.

Are Mathematical Truths Necessary?

The puzzle of existence raises a profound question: How deep do explanations go? Does the world's existence have a basis of explanation that cannot itself be explained, or does explanation extend all the way to the foundation? There is something alluring about the notion that explanation might run so deep that it leaves no major questions unanswered. The prospect that mathematics offers necessary truths makes it intriguing as a potential source of ultimate explanation. If mathematical explanations show why things could not be other than they are, then a mathematical answer to the puzzle of existence might reveal why there had to be a physical world. We would have explanation all the way down. But does mathematics really offer that kind of necessary truth?

69 Balaguer, *Platonism and Anti-Platonism in Mathematics*, 6.

70 Balaguer, *Platonism and Anti-Platonism in Mathematics*, 48–49.

Initially at least, it appears to many that it does. It is hard to think of anything more necessary, certain, and unalterable than the things we take for granted about arithmetic, geometry, and other mathematical subjects. Indeed, for thousands of years, it was taken for granted that mathematics was an arena where there was only one way that things could be. However, since the nineteenth century, developments from within mathematics itself have provided reasons to question that assumption.

Over two millennia earlier (around 300 BCE), the Greek mathematician Euclid composed a treatise, *The Elements*, establishing a system of plane and solid geometry. From fixed premises, Euclid reasoned to propositions that necessarily followed. The methodology is based on the idea that if one begins with indubitable truths and each step in the reasoning from one proposition to another is warranted, then the conclusions must be true. Euclid's starting points include both definitions and axioms. The definitions simply establish the meaning of terms, and the axioms are statements considered so basic and self-evident as to be beyond question. The system proved extraordinarily powerful. Based on its premises, one could demonstrate innumerable propositions, including, for example, the Pythagorean Theorem (showing that summing the squares of a right triangle's two shorter sides yields the square of the longer side). Euclid's geometry was not merely a handy tool for solving practical problems but was thought to reveal the very nature of the physical world. A system so elegant and useful had to be telling us about the cosmos itself. Since geometry concerned extended figures and shapes, it described the true nature of space.[71] For millennia, it was not necessary to specify that you were doing "Euclidean" geometry, because that was the only way to do it.

The modern discovery that disrupted mathematics centered on the last of Euclid's five axioms. In brief, the first four axioms set forth the following: that a line can be drawn between any two points; that a line can be indefinitely extended; that a circle can be drawn around any point with a given radius; and that all right angles are ninety degrees. The fifth axiom, known as the "parallel postulate," states that given a line and a point not on the line, you can draw one and only one new line through the point that never intersects with the first line. That is, you can draw exactly one line through the point that is parallel to the first line.

While Euclid's system worked beautifully and no one questioned the truth of the axioms, the parallel postulate did not seem as obvious or self-evident as the other four axioms, and many mathematicians suspected that it could be deduced from them. Prizing efficiency, mathematicians prefer a

71 Barrow, *Pi in the Sky*, 9.

more elegant system that can prove the same number of propositions with fewer assumptions. Thus, it would have been a notable advance to reduce the axioms in Euclid's system from five to four. That quest, however, went unfulfilled, as no one could prove the parallel postulate from the other axioms. Although frustrating, the lack of a proof did not in itself present a serious problem, because everyone accepted the parallel postulate regardless. The crisis came in the middle of the nineteenth century when three mathematicians working independently from one another—János Bolyai, Carl Friedrich Gauss, and Nikolai Lobachevsky—showed that it was possible to construct alternative geometries that rejected the parallel postulate. On certain curved surfaces—something with the shape of a saddle, for instance—there is not only one parallel line that can be drawn, but infinitely many. It turned out that the parallel postulate only applied to flat surfaces, while the new "non-Euclidean" geometries—without the parallel postulate—were appropriate for curved surfaces. Like Euclidean geometry, these alternative systems were self-consistent, elegant, and powerful. There was not only one way to do geometry after all.

The discovery of non-Euclidean geometries had a tremendous impact on the way that people thought about the relation between mathematics and the world. As long as there was only one geometrical system—and, thus, only one set of geometrical truths—it seemed reasonable to assume that geometrical propositions directly inform us about the physical space in which we live. However, the development of non-Euclidean geometries called that assumption into doubt. If conflicting propositions can be demonstrated in different self-consistent systems, then it is no longer possible to draw a one-to-one correspondence between mathematics and the world around us. Once you have more than one powerful system of geometry, how can you know which one describes the cosmos? Theorems based on Euclidean axioms are only true in Euclidean spaces (that is, with respect to structures where Euclidean axioms hold). The actual nature of the space in our physical reality became an empirical question. Instead of relying on deductive proofs, you have to look and see. Indeed, less than a century after Bolyai, Gauss, and Lobachevsky exploded the myth that there was only one right way to do geometry, Einstein's theory of general relativity showed that non-Euclidean geometries described the physical world more accurately than did Euclid's.

The discovery of non-Euclidean geometries had even broader ramifications, highlighting that mathematical claims were only true relative to a particular system.[72] No set of axioms was the uniquely correct one. Since a wide variety of axioms could be adopted, and different axioms imply

72 Barrow, *Pi in the Sky*, 11–13.

different propositions, it became essential to focus on the specific axioms to be applied within a given mathematical system.

Systems are crucial in mathematics, because we need fixed starting points in order to show how they entail certain conclusions. Given a specified set of axioms and rules for employing them, mathematical statements in a chain of reasoning follow necessarily from one another. Many people are familiar with using mathematical systems from the geometry that they studied in school. Based on certain statements taken to be true, we can derive other propositions (theorems). Axiomatic systems are also familiar from arithmetic, which today is commonly conducted in accordance with the so-called Peano axioms.[73] Along with alternative systems of arithmetic, researchers have developed numerous complex systems for other mathematical subjects, such as topology.[74]

The impact that axiom selection has on the propositions that are true within a system raises questions about the nature of mathematics. The leeway for the axioms to vary may call into question the necessity of mathematical truths. If the axioms for a particular system are not rooted in necessity—but instead depend on choices made by the people constructing a particular system—then what does this mean for the character of the truths implied within the system?[75] Does it suggest that there is an unavoidably arbitrary element lying at the foundations of mathematics?

The question has critical importance for a theory like Tegmark's, which proposes a mathematical answer to the puzzle of existence. The mathematical universe, by Tegmark's lights, is fully necessary from beginning to end. Tegmark's theory posits a necessary link between mathematics and the world, with every mathematical structure existing as an external reality, which is why it posits that our universe is only one of countless structures within the multiverse. While the multiverse comprises a tremendous number of distinct universes, there is only one way that reality as a whole—encompassing all of the different universes—can be. Thus, the theory only works if the multiverse's composition is fully independent of our own thought. Indeed, Tegmark defends the MUH by analyzing what it means for a structure to exist apart from any cultural baggage. Moreover, in advancing reasons to embrace the MUH, Tegmark relies on the supposed necessary character of mathematics. For instance, he contends that only a

73 Russell Marcus and Mark McEvoy, *An Historical Introduction to the Philosophy of Mathematics: A Reader* (Bloomsbury, 2016), xxiii; Hamlin, "Toward a Theory of Universes," 573. The Peano axioms are named in honor of Giuseppe Peano, who axiomatized a system for the natural numbers in the late nineteenth century.

74 Marcus and McEvoy, *An Historical Introduction to the Philosophy of Mathematics*, xxiv.

75 Marcus and McEvoy, *An Historical Introduction to the Philosophy of Mathematics*, xxv–xxvi.

mathematical theory of existence can explain the properties that things have, because other kinds of explanations fall prey to an infinite regress.

But does the discovery of non-Euclidean geometries pose a genuine threat to the necessity of mathematics? Not all mathematicians think that it does. Since Bolyai, Gauss, and Lobachevsky discovered non-Euclidean geometries, many mathematicians have sought to understand mathematics in ways that leave the necessary character of its truths undisturbed. They argue that the variability of the axioms in different systems does not threaten the necessity of mathematical truths. Now, no one today can deny the role that human choice plays with respect to the axioms to be used in a particular system by mathematicians at a particular time. However, this recognition can be reconciled with the necessity of mathematical truths if one believes—as many mathematicians do—that there is necessity underlying what counts as a self-consistent set of axioms. That is, if all of the available sets of axioms are necessarily determined, then there remains a crucial sense in which the propositions entailed by any particular set of axioms are necessarily true. Instead of saying that a proposition is always and everywhere true, we may simply say that a proposition is necessarily true given a particular set of axioms. With this adjustment in the framing of mathematical claims, mathematical propositions remain necessarily true, because necessity underlies the options for axiomatic systems and the truth of propositions within each system.

This way of understanding the role of axioms is consistent with Tegmark's Platonist claim that every structure described by a mathematical system exists independently of physical things or ideas conceived by human minds. Tegmark maintains that the nature of mathematical systems specifies the forms that they can take and, thus, that there is a fixed set of formal systems waiting to be discovered. As Tegmark writes,

> Although whims of human fashion influence the choice of which particular formal systems we explore at any one time, and which aspects thereof, we are continually increasing the amount of charted territory. The street map of mathematical structures is 'out there,' and any intelligent entity who begins to study any corner of it will inevitably discover at least the main plazas and connecting boulevards, even if many charming back alleys and sprawling suburbs are missed due to cultural prejudice.[76]

However, even if non-Euclidean geometries do not undermine the necessity of mathematics, another development about a century later—Kurt

76 Tegmark, "On Math, Matter, and Mind," 769.

Gödel's "incompleteness theorems"—may be seen as problematic for a mathematical theory of existence like Tegmark's. Gödel's theorems cast doubt on the hopes and expectations of many observers regarding mathematical knowledge. Non-Euclidean geometries had given rise to a greater focus on the selection of the governing axioms and rules for a particular mathematical system. Conceiving mathematics increasingly as the construction of formal systems, many mathematicians focused on developing internally consistent systems. Even if mathematics did not establish unconditional truths about the world, it could nevertheless enable us to reliably decide the truth of statements made within an axiomatic system. Championing this approach, mathematician and philosopher David Hilbert in 1920 announced the aspiration to develop a basis for mathematics with a single set of axioms. If successful, the program would enable mathematicians to determine through step-by-step procedures whether any claim made within the system was true. This ambitious aim was to be pursued through the construction of an axiomatic system with its own precisely defined formal language and rules for producing well-formed statements in that language. Using the system's rules properly would guarantee the avoidance of self-contradictions. Moreover, Hilbert believed that the system would be able to prove its own consistency. Thus, Hilbert's dream was to develop a single system that could determine the truth or falsity of all claims while demonstrating its own validity. He invited the greatest mathematicians to join in this project.[77]

Unfortunately, not only was the project not completed, but in 1931 the mathematician and logician Kurt Gödel (1906–78) announced his two incompleteness theorems, which showed that it never could be.[78] The incompleteness theorems revealed inherent limitations of axiomatic systems.[79] The first of Gödel's incompleteness theorems proved that any internally self-consistent axiomatic system would give rise to statements whose truth could not be determined by proofs within the system.[80] Worse, this leak in the boat could not be plugged by adding more axioms, because the new axioms would simply give rise to new undecidable statements. The second incompleteness theorem undermined another vital part of Hilbert's project: it showed that a formal system could not verify its own consistency. The force of the second theorem can be illustrated in everyday terms with reference to a standard calculator. We use calculators to add, subtract, and

77 Barrow, *Pi in the Sky*, 19.

78 Barrow, *The Universe That Discovered Itself*, 288.

79 More specifically, the theorems apply to mathematical systems that are strong enough to represent arithmetic, which include the kinds of theories that were the focus of Hilbert's program.

80 David Deutsch, *The Fabric of Reality: The Science of Parallel Universes—and Its Implications* (Penguin Books, 1997), 235.

carry out many other computations. Suppose that in the middle of conducting a particular calculation, we begin to wonder about the reliability of the calculator's results. How might we confirm their accuracy? Gödel's second theorem demonstrates that we cannot possibly use the calculator itself to prove the correctness of its results. If we want to ensure the calculator's trustworthiness, we must check its results against mental calculations or the outputs of another, more powerful calculator. Together, the two theorems showed the impossibility of fulfilling Hilbert's dream. Since Gödel supported the theorems with logically rigorous proofs, the theorems did not merely point out deficiencies in the power of modern mathematics; they demonstrated limitations on what was logically possible.

Gödel's theorems pose a challenge that is distinct from that posed by the discovery of non-Euclidean geometries. Here, the difficulty concerns not the appearance of arbitrariness in establishing a system, but, rather, the potential for indeterminacy and uncertainty regarding the truth of mathematical propositions within a system. As we have seen, Tegmark hypothesizes a multiverse in which each universe is a mathematical structure. Every mathematical structure actually exists as a universe, and, thus, there is no gap between possibility and reality. This places tremendous importance on what the MUH recognizes as a mathematical structure. Since the MUH equates mathematics and reality, it would be no less absurd to suggest that mathematics contains self-contradictions than that reality itself does. Now, recall that for Tegmark a mathematical structure is described by a quantitative expression within a well-defined formal system that makes clear how its axioms and rules can be used to demonstrate the truth of statements made within the system. Given the various premises that underlie the MUH, the incompleteness theorems are potentially problematic, because they identify limitations on the ability to definitively determine the truth of claims in a system, as well as on a system's ability to prove the consistency and correctness of the conclusions it generates. As Tegmark writes, "Gödel's work might make us worry that the MUH makes no sense ... because our Universe would be somehow inconsistent or undefined."[81]

Tegmark does not ignore or downplay the questions raised by the incompleteness theorems. He acknowledges the potential difficulties that the theorems pose for his theory, and shares thoughts on how those difficulties might be addressed, without purporting to have yet arrived at complete solutions.[82] After recognizing "the objection involving Gödel to be

81 Tegmark, *Our Mathematical Universe*, 331.

82 Tegmark, "The Mathematical Universe," 35–39; Tegmark, *Our Mathematical Universe*, 330–35; Tegmark, "On Math, Matter, and Mind," 788.

very interesting and subtle,"[83] he notes that a core problem that the incompleteness theorems raise concerns which structures are "sufficiently well defined" to "qualify for membership" in the multiverse.[84] The core of the issue concerns the standards that establish what counts as a mathematical structure, and, thus, determine which structures are granted admittance as actually existing universes within the cosmic multiverse. Rather than trying to explain away the challenge posed by Gödel's theorems, Tegmark frames his account of what counts as a structure around it. Conceding that undecidability poses a potential problem for his theory, he adopts decidability as a constraint on reality. In doing so, he makes his response to the threat posed by Gödel a key part of the MUH itself. Tegmark hypothesizes that "only Gödel-complete (fully decidable) mathematical structures have physical existence."[85] This means that the multiverse only includes structures whose underlying mathematical systems allow for the determination of what qualifies as a valid theorem within the system.

In responding to the challenge posed by the incompleteness theorems, Tegmark introduces a specific restriction on what counts as a mathematical structure, which he calls the Computable Universe Hypothesis (CUH). According to the CUH, the relations constituting a structure must be computable functions, meaning that the structure could be produced by a program through a finite number of computational steps.[86] We have noted the significance that Tegmark accords to the idea of computation. For instance, he views an objective description as one that could be processed by a computer, and he holds that a supercomputer could calculate all physical behavior in our universe based on the theory of everything.[87] Indeed, about five years after Gödel announced the incompleteness theorems, the mathematicians Alonzo Church and Alan Turing demonstrated critical connections between decidability in mathematics and computability, ideas which can largely be translated from the language of one context into the other.[88] The requirements of a computable function include that its instructions can be specified in a finite length and that its procedures can be completed in a finite time period. While the MUH includes an infinite number of

83 Tegmark, "On Math, Matter, and Mind," 788.
84 Tegmark, *Our Mathematical Universe*, 330.
85 Tegmark, "On Math, Matter, and Mind," 788.
86 Tegmark, "The Mathematical Universe," 129.
87 Tegmark, "The Mathematical Universe," 102.
88 Panu Raatikainen, "Gödel's Incompleteness Theorems," in *The Stanford Encyclopedia of Philosophy*, ed. Edward N. Zalta (2022), § 1.3, accessed December 10, 2024, https://plato.stanford.edu/archives/spr2022/entries/goedel-incompleteness/. As discussed further below, Tegmark's restriction of the mathematical universe to Gödel-complete structures or systems entails the potentially problematic result of ruling out many subjects widely accepted today as legitimate parts of mathematical study.

universes, the CUH requires that all of them must be "defined by computable functions."[89]

The introduction of the CUH underlies Tegmark's claim that "no physical aspects of our universe ... are uncomputable/undecidable, eliminating the ... concern that Gödel's work makes it somehow incomplete or inconsistent."[90] He thinks this must be the case because, in his own words, "I don't know exactly what properties our physical reality has, but I'm confident that these properties exist in the sense of being well defined: I have no doubt that nature knows what it's doing!"[91]

While presenting the CUH as a response to the incompleteness theorems, Tegmark acknowledges that it may introduce its own challenges for the mathematical universe. One potential difficulty concerns a problem in computer science—known as the "halting problem"—first brought to light by Church and Turing, working independently during the 1930s. Closely related to Gödel's proof that some theorems in mathematics are undecidable, the halting problem recognizes that there is no algorithm that can reliably determine for all computer programs whether they will be able to complete their instructions in a finite length of time. In other words, we cannot always know whether a particular computer function is computable. Like the incompleteness theorems, the halting problem raises potential problems for Tegmark's theory by introducing indeterminacy regarding what counts as a well-defined mathematical structure that merits admittance to the multiverse.[92]

In response to these concerns regarding indeterminacy, Tegmark restricts what the MUH recognizes as a well-defined structure. In particular, the CUH allows only structures described by a function that *is* computable. However, as Tegmark recognizes, the CUH imposes "extremely restrictive" conditions on what counts as a mathematical structure, and, thus, "poses serious challenges that need to be resolved."[93] A particularly significant challenge is that the CUH rules out many structures that mathematicians today generally accept as parts of the discipline. Tegmark does not shy away from this implication, surmising that many things currently viewed as legitimate subjects of mathematical study will turn out to be "a mere illusion, fundamentally undefined and simply not existing in any meaningful sense."[94] For example, the CUH would rule out "one of the most popular mathematical structures of our time: that of the so-called real numbers," including pi,

89 Tegmark, *Our Mathematical Universe*, 332.
90 Tegmark, "The Mathematical Universe," 136.
91 Tegmark, *Our Mathematical Universe*, 333.
92 Tegmark, *Our Mathematical Universe*, 330–31.
93 Tegmark, *Our Mathematical Universe*, 333.
94 Tegmark, "The Mathematical Universe," 136.

"whose decimals go on forever," because it would require infinite information to process them.[95] For similar reasons—namely, that the CUH cannot accommodate real numbers because they "require infinitely many bits to specify[96]—the CUH also contradicts many aspects of contemporary physics. As Tegmark notes, "virtually all historically successful theories of physics violate the CUH, and ... it is far from obvious whether a viable computable alternative exists."[97]

Tegmark tentatively proposes ways of avoiding the apparent inconsistency between the CUH and contemporary physics. One of his most notable suggestions is to explore "abandoning the continuum [the idea that the physical world divides into infinitely small parts] and trying to recover it as an approximation." In support of the idea, he notes that physicists have "never measured anything ... to more than about 16 significant digits, and no experiment has been carried out whose outcome depends on the hypothesis that a true continuum exists, or hinges on nature computing something uncomputable."[98]

While some may reasonably consider it a strike against Tegmark's theory that it requires us to reject so many prevailing views about mathematics and the world, it should be noted that some of the greatest advances in human understanding have overthrown beliefs long taken for granted. At any rate, it is to Tegmark's credit that he openly acknowledges potential challenges for this theory. By so transparently recognizing objections and offering what he considers promising directions for addressing them, Tegmark makes it easier for other experts to join the continuing debate over the implications of technical questions in mathematics and physics for the quest to explain the world's existence.

Living in the Mathematical Universe

The previous two sections have discussed challenges for a mathematical explanation of the world's existence. At this stage, however, all of the major approaches to explaining the physical world's existence face substantial challenges, and it remains too early to tell if those facing a mathematical approach will be overcome. In this section, we turn to a different set of questions concerning the implications of a mathematical cosmos. Assuming that

95 Tegmark, *Our Mathematical Universe*, 331.
96 Tegmark, "The Mathematical Universe," 136.
97 Tegmark, "The Mathematical Universe," 138.
98 Tegmark, *Our Mathematical Universe*, 330–31.

a mathematical theory like Tegmark's is correct, what would it mean for the nature of reality and our place in it?

In one significant respect, the implications of a mathematical approach resonate with those we discussed in the last chapter pertaining to a scientific approach. In particular, both vantage points for attacking the puzzle of existence refuse to accord human beings an inherently special place in the cosmic order. The scientifically observed behavioral regularities that govern things in the world are thoroughly egalitarian. The principles of physics apply to the constituents of different objects, and all objects are ultimately composed of the same constituents. The natural laws abide no exceptions. From a purely mathematical standpoint, too, human beings can expect no special seat at the table. In the mathematical universe, human beings, like everything else, are simply manifestations of a mathematical structure's properties. Neither the scientific nor mathematical universes have any place for hierarchy; they do not designate some sorts of things as inherently more significant than others.

Scientific and mathematical approaches to the puzzle of existence each operate in what we might call a one-track manner. Whether through behavioral regularities in the one case, or the necessity of mathematical relationships in the other, both the scientific and mathematical universes treat everything within their jurisdiction on equal terms, which is to say with the same indifference. The ideas of purpose, meaning, and narrative, therefore, have no purchase. To the extent that particular human beings experience such ideas as real, they come into play only from the perspective of those human beings themselves.

Tegmark cheerfully embraces the anti-hierarchical character of the mathematical universe. In the MUH, human beings, like everything else, are simply elements within a larger mathematical structure.[99] Even the parts of ourselves that might seem most exceptional do not stand distinct from the mathematical cosmos, within which we are fully embedded. In Tegmark's words, even the human mind "emerges from math, as a self-aware substructure of an extremely complicated mathematical structure."[100] Reality's mathematically determined nature applies fully to our own thoughts, as Tegmark affirms that "an infinitely intelligent mathematician could in principle deduce the properties of both its material content and the minds."[101]

In emphasizing that the mathematical universe lacks hierarchy, Tegmark characterizes his theory as the ultimate extension of Copernicanism. Before the publication of Copernicus' *On the Revolutions of the Celestial Spheres* in

99 Tegmark, *Our Mathematical Universe*, 271.
100 Tegmark, "On Math, Matter, and Mind," 779.
101 Tegmark, "On Math, Matter, and Mind," 779.

the 1540s, it was widely accepted that the Sun and other celestial bodies revolved around the Earth. Although in a society dominated by a Christian worldview this certainly did not mean that our material home was the best or highest place one could attain, it did mean that the Earth occupied a special place, lying at the center of the cosmos. Copernicus' work was revolutionary because it provided overwhelming evidence that the Sun did not orbit the Earth; it was the other way around. Since Copernicus' day, additional discoveries have continued to chip away at the idea that the Earth occupied any kind of special status. Later, cosmologists learned that our solar system was only one of many in the galaxy. Today we know that there are hundreds of billions more galaxies in the observable universe. If Tegmark's theory is right, it effects an even more dramatic demotion. We do not just live in *the universe* but, rather, in a *multiverse* that contains countless different universes. Our universe's particular characteristics—including its hospitality to life and organisms like ourselves—are in no sense the product of cosmic purposes, plans, or narratives featuring us as main characters. Rather, our universe, like all universes, is just a mathematical structure, and all mathematical structures exist as universes. The particular features of our universe have not singled us out for exceptional treatment. They are distinctive simply because each mathematical structure gets a room in the multiverse hotel.

Notwithstanding the resonance between scientific and mathematical approaches in their refusal to accord human beings a special place, they also have radically different implications due to their disparate foundations. The most basic truth in the scientific universe is that objects exist with particular physical properties, and science cannot identify the reasons that entities exist or have the particular properties they do. Since we cannot identify necessity at the ground floor, the only way we can learn about the world scientifically is through empirical observations. The discipline aims to summarize the ways that physical things behave and interact, with all theories subject to revision if new observations conflict with old understandings. By contrast, if a mathematical solution to the puzzle of existence is successful, it means that reality is necessary from head to toe. Mathematical truths cannot be other than they are, so reality cannot be other than it is. Even the foundation of explanation is explained, because all truths are rooted in the relations among mathematical objects. Unlike the scientific universe, nothing in the mathematical universe is chancy or uncertain. Everything is fully determined, and there are no alternative scenarios.

To further highlight the divide between scientific and mathematical approaches, take an issue we considered in discussing the historical roots of inquiry into ultimate questions: the nature of change and continuity.[102] Recall the diametrically opposed positions of Heraclitus and Parmenides, with the former viewing change as pervasive and the latter denying change any place in reality at all. The debate remains as salient for us today as it was for the ancients. The spirit of Heraclitus' thought is very much alive in today's science. What is science if not the study of dynamism? Testing predictions is central to science because researchers aim to identify the factors that are responsible for change. When we view the world through a scientific prism, we do not see alteration as an illusion or transgression against what is most real. To the contrary, the most basic truths describe the manner in which things change. Democritus and other Atomists were prophetic in claiming that the things around us were composed of tiny bits that we could not see. Nevertheless, we no longer believe these bits are indestructible or unchanging. Contemporary physics does not study entities immune from transformation but quantum fields aboil with the possibility of random fluctuations. As the Heisenberg uncertainty principle insists, determinate stability is just not on offer in the scientific universe.

On the other hand, the central feature in a mathematical approach to the puzzle of existence is its fixed necessity. The spirit of Parmenides is evoked by Tegmark's denial that change is real. In the mathematical universe, the flowing of time itself is merely an illusion.[103] As Tegmark writes, "Our external physical reality is an abstract, immutable entity existing outside of space and time."[104] Unlike the restless energy relations described by quantum physics, the ratio of a circle's circumference to its radius allows for no updates or modifications.

The mathematical universe also has surprising implications for the nature of our knowledge about the world. Tegmark's ultimate Copernicanism means that discoveries about the universe do not tell us about the nature of reality as a whole, since our world's properties are merely an "environmental accident."[105] According to the MUH, the mathematical formula describing our universe does not reveal anything "fundamental about reality, but instead which particular mathematical structure we happen to inhabit, like a multiversal telephone number."[106] The reference to a structure we "happen to inhabit" requires clarification, since that phrasing normally suggests the notion of chance. But there is no such thing as chance in the mathematical multiverse. Every structure exists as a world in the multiverse and could not

102 See Chapter Two.
103 Tegmark, "The Mathematical Universe," 118.
104 Tegmark, *Our Mathematical Universe*, 274.
105 Tegmark, "The Mathematical Universe," 117.
106 Tegmark, "The Mathematical Universe," 117.

be other than it is. However, since we only live in one of the structures that comprise the multiverse, our universe's features are not the rules that apply throughout the multiverse. Every mathematical structure exists as an external reality, and we reside in just one of them.

The implications of the mathematical universe for our ability to understand reality are dramatically different in the short-run than the long run. In the near term, our understandings are quite limited. We do not yet even know the "theory of everything" for our universe, much less for the other universes in the multiverse. Worse, even if we knew the theory of everything for each universe, we presently have little idea how to derive a universe's features from the mathematical formula that describes it. At the same time, the mathematical cosmos has stunningly optimistic implications for our long-run prospects. Since we can reason about mathematics, there is in principle nothing to stop us from deductively unlocking the full range of truths. Thus, the MUH suggests a future in which human beings have a comprehensive understanding of the cosmos, including why each universe has the properties it does and why we experience the world as we do.

Apart from the knowledge we might possess, an unmistakable implication of the mathematical universe is that there is an enormous gulf between the way things appear to us as we move through our lives and reality's true nature. We certainly experience our lives as though time flows, objects change, and events occur one way when they might have occurred another. Yet the mathematical universe rejects these beliefs.

A theory making such counterintuitive claims owes us an explanation of why the truth is so well-disguised. Tegmark's positing of a divergence between reality and the way that the world appears to our senses brings Plato to mind, although his approach to explaining the gap is quite different.[107] Plato famously relies on a dualistic worldview that distinguishes the ever-fluctuating particular things perceived by our senses from the unchanging, perfect forms that lay at the foundation of reality. Since Tegmark's MUH comprises only mathematical objects, his explanation for our life of illusion cannot be dualistic in that manner. For Tegmark, the crucial distinction is not one between particulars and forms, but, rather, between subjective and objective perspectives. An objective perspective is one that apprehends the whole all at once and, thus, enables one to see reality as it is. On the other hand, a subjective perspective generates illusions because it is restricted to a particular vantage point.

To illustrate the distinction, Tegmark uses the metaphor of a frog and a bird. From the frog's viewpoint at a particular place in the mathematical

107 See the discussion of Plato in Chapter Two.

structure, it experiences the impression of flowing time.[108] Soaring high above in the sky, however, the bird enjoys the objective view of one studying an entire mathematical structure. This vista enables the bird to see the universe as it really is: constituted of fixed and unalterable abstract relations that stand outside time. In this vein, Tegmark asks us to consider

> a world made up of pointlike particles moving around in three-dimensional space. In four-dimensional spacetime—the bird perspective—these particle trajectories resemble a tangle of spaghetti. If the frog sees a particle moving with constant velocity, the bird sees a straight strand of uncooked spaghetti. If the frog sees a pair of orbiting particles, the bird sees two spaghetti strands intertwined like a double helix. To the frog, the world is described by Newton's laws of motion and gravitation. To the bird, it is described by the geometry of the pasta—a mathematical structure.[109]

Shifting to a cinematic metaphor, Tegmark compares subjectivity to experiencing one frame from a film reel at a time while objective reality comprises the entire movie all at once.[110]

Tegmark also employs a perspectival strategy to explain the illusion of chance. As we have seen, the dominant interpretation of quantum theory maintains that randomness is real.[111] To explain why randomness appears real to physicists, Tegmark adopts a competing account of quantum theory: the Many Worlds Interpretation. Adopting this alternative interpretation allows Tegmark to once again draw a contrast between two perspectives. Here, he distinguishes the subjective perspective of an individual from a scientist's objective perspective. An individual lives and sees things in only a single world, but the scientific vantage point allows us to recognize that every outcome allowed by the wave function is realized in a distinct parallel world.

According to the Many Worlds Interpretation, when a quantum branching occurs, I only perceive one of the resulting worlds: the particular one in which I reside. Yet, there is another version of me in every branched universe, and each one is suffering under the same misimpression. From the limited perspective of one branched world, it appears that things could have turned out differently. It is only from the perspective of the whole that we can recognize the truth that contingency is merely an illusion. The objective

---

108 Tegmark, "The Mathematical Universe," 106.

109 Tegmark, "The Mathematical Universe," 106.

110 Tegmark, "On Math, Matter, and Mind," 770.

111 See the discussion of quantum theory—and especially the Copenhagen interpretation—in Chapter Three.

reality is that the full set of possibilities is realized in the only way that it could be. The frog, stuck in only one world, thinks randomness is real. The bird, able to see the totality, understands that randomness is an illusion and everything is determined.[112] The mathematical cosmos, then, is one in which things are not what they seem, and our initial impressions about the world are likely to be wildly off base. Nevertheless, with the aid of the objective perspective offered by mathematics, all reality's secrets might one day be open to our understanding.

112 It is important to note that the Many Worlds Interpretation of quantum physics is very different from Tegmark's mathematical multiverse. Tegmark's theory holds that we are located within one of countless universes in the mathematical universe, manifesting different physical laws. Like other contemporary physicists, Tegmark accepts the quantitative formulas furnished by quantum physics. However, as discussed in the previous chapter, physicists disagree over how to interpret quantum physics. Tegmark endorses the Many Worlds Interpretation, which holds that there is no collapse of the wave function. Instead, every possible state of a quantum is realized in a distinct world, eliminating the distinction between possibility and actuality. Although different states are realized in each of the branching worlds, they all nevertheless have the same physical laws. Thus, they do not constitute separate universes within Tegmark's mathematical multiverse. For Tegmark, then, everything described by the MWI happens within our particular universe.

Questions for reflection and discussion

1. What are the most significant differences between the fields of science and mathematics?
2. What best explains the "unreasonable effectiveness of mathematics in the natural sciences"?
3. What is the best reason to take mathematical theories of existence seriously?
4. How persuasive is Tegmark's explanation of why mathematics is so useful in the sciences?
5. Do you think physicists will ever uncover a single formula that predicts all behavior?
6. How plausible do you find Tegmark's idea that the cosmos is a multiverse of diverse worlds?
7. What is the most convincing view regarding the nature of mathematical objects?
8. Can the Platonist view on mathematical objects explain how we learn about mathematics?
9. Do you think that non-Euclidean geometries pose a problem for mathematical theories of existence?
10. What does a mathematical explanation of the world imply about human nature?
11. What does Tegmark's theory suggest about our ability to understand the world?

5.

The Good Universe

The Distinctiveness of Value-Based Explanations

It is common to think of value as secondary. The world came first, many assume, and value only appeared at some later point in the story. In this view, tangible entities preceded any truths about what is good or bad. The physical had reality before the evaluative. Those holding this view do not necessarily deny the reality of value, but they insist that it is dependent on other things in the world (such as living organisms, for whom one outcome might be considered better than another). This chapter examines an approach to the puzzle of existence that reverses the usual order. As a bumper sticker, this approach would declare "Value First, then the World." This approach to explaining the world's existence, which we will refer to as "axiological" (from the Greek term, *axía,* meaning "value" or "worth"), posits that value has an existence independently of the physical world, for which it is responsible. To the question of why anything physical exists, it replies: because it is good that it does.

Explanations based in value or goodness (terms used interchangeably here) are familiar to us in our daily lives. Why did you take that trip? Why did a friend choose one job over another? What made watching the movie fun? These are not the sorts of questions likely to be answered in a satisfying way by scientific theories or mathematical proofs. Instead, the reasons will speak in the language of value. Perhaps you took the trip to enjoy good food and music, the friend chose the job to gain experience in a field she loves, and the movie's stunning visuals made it a pleasure to behold.

We use the term "value" here in its broadest sense, encompassing all of the reasons things might matter to us, including, for example, because they are beautiful, useful, funny, inspirational, or joyous. People sometimes refer to value in a narrower sense, as when they use the term "value judgment" to indicate condemnatory attitudes toward immoral behavior. Thus, it is worth stressing that we will be employing the idea of value more expansively to include all the different qualities that might make things meritorious, admirable, or worthwhile. Since we are interested in attempts to ground existence in any kind of value, we need not establish definitions specifying the boundaries between various value categories. For our purposes, the crucial distinction is between explanations rooted in value and all other explanations. In particular, our focus here is on explanations for the existence of the physical world that are based most fundamentally in value or goodness.

It is important at the outset to distinguish axiological answers from those based in theism (discussed in the next chapter). Over the last two millennia, the Abrahamic religions—including Judaism, Christianity, and Islam—rose in prevalence throughout much of the world. As a result, discussion about the cosmic significance of goodness has taken place largely within a theistic context. The Abrahamic religions generally posit an all-powerful God who was responsible for creating the world. In these religions, God is often described as a perfectly good divine being who created the world because of the goodness that it would embody. To note a well-known example, the Hebrew Bible opens by recounting that God created the world over six days and saw that it was good.[1] The idea that God is the explanation for the world's existence is thus more familiar to many people than the idea that goodness itself is the explanation.

The common association between deity and goodness makes it easy to confuse theistic and axiological theories of existence. However, despite the significant role that goodness plays in certain theistic frameworks, from a theistic vantage point it is God—a divine personal being—who lies at the foundation of reality and is ultimately responsible for the world's existence. By contrast, in an axiological theory, goodness itself is the foundation that bears responsibility for the world's existence.

The ideas of deity and goodness do not mean the same things, and they are in principle separable. One might posit a powerful supernatural being who is not perfectly good. (In this vein, the gods of ancient Greek religions engaged in many actions that were anything but praiseworthy.) Conversely, one might find the source of existence in goodness without attaching it to a supernatural being. That said, an axiological approach is not necessarily hostile to or inconsistent with theism, and the two may be integrated. For

1 Genesis 1:31.

instance, one might endorse theism while maintaining that the goodness of God's being is the reason why God exists in the first place. Nevertheless, even if some may find a congenial relation between axiology and theism, they represent different views about what is most fundamental.

In this chapter we are using the term "axiological theory" (or "axiological approach" to the puzzle of existence) in a specific sense: to describe accounts holding both that value has an existence that is independent of the physical world, and that value itself explains the world's existence. It is important to highlight the way that we are using the term here because many value theorists, while taking value seriously as a basis for some kinds of explanations, nevertheless view value as dependent on—or derivative from—physical existence or something other than value. However, a theory that views value as ultimately based in the physical world is ill-suited to answering why the physical world exists, because it would be assuming the very fact that it is supposed to explain. Understood in the manner we are using the term here, axiological theories differ profoundly from either scientific or mathematical explanations for the world's existence. Consequently, arguments for axiological theories have a distinctive texture from anything that we have seen thus far.

In seeking to provide reasons for things being the way they are, science relies on observable physical regularities, while mathematics looks to unalterable truths about certain necessary relationships. Neither field can give us an axiological reason for the world's existence. Science alone is not well-suited because propositions about values do not generate predictions that can be confirmed with empirical measurements of specified interactions. Mathematics is likewise inapt because we do not generally limit our thinking about value to the deduction of demonstrably necessary truths from value-neutral axioms.[2] In any event, most practitioners of both science and mathematics consider it essential to their respective disciplines that they not rely on beliefs about value or goodness to establish their claims.

At the same time, there is no reason why an axiological theory must depreciate the significance of science and mathematics as disciplines and useful modes of explanation. Generally speaking, there need be no conflict between science and mathematics, on the one hand, and an axiological approach to the puzzle of existence, on the other, because they may be viewed as explaining different things. However, when it comes to asking why there is a physical world, an axiological theory maintains that the answer is to be found in value, which does not depend on anything physical for its existence.

2 For example, few philosophers today would attempt to resurrect Baruch Spinoza's dream of a wholly deductive ethical system modeled on Euclid's geometry. While some philosophers have attempted to axiomatize their ethical theories in order to deduce a set of necessary ethical truths, in those cases the axioms themselves make claims about value.

An axiological approach, then, maintains that we can know things beyond what can be shown through laboratory experiments, empirical observations, formal logical arguments, or deductive proofs. As we will see, rather than purporting to present a definitive demonstration, axiological arguments tend to be multi-layered and comparative, contending that, all things considered, value offers the most compelling explanation for the existence of the world in which we find ourselves.

The Pervasiveness of Value

Although arguments for an axiological theory are not scientific in the sense of making claims that give rise to falsifiable predictions, neither do they purport to derive purely from propositions that we can know just through rational reflection. Rather, they often appeal to broad observations about the nature of the world. Most notably, teleological arguments are based in the evident existence of certain good things in the world that seem difficult to explain without making assumptions about value's foundational significance. Before turning to teleological arguments in the next section, this section considers observations about the pervasive role that value plays in our lives. To be sure, recognizing the pervasive role of value in human lives does not make a definitive case that goodness itself is the reason for the world's existence. The aim here, however, is simply to provide additional reasons why one might find the idea that value or goodness plays a fundamental explanatory role at least initially intriguing.

For human beings, value does not come into the picture downstream, but lies at the font of our experience. We do not even have the option of encountering the world in a way that is value neutral. Of course, our experience of the world as deeply value-laden does not, in itself, establish that value has an independent existence. Much less does it establish that value is the source of ultimate explanation. Nevertheless, axiological theories have an advantage in being particularly well-positioned to make sense of the way that value permeates our interactions with the things around us.

While things in the natural world were commonly conceived of in ancient and medieval times as having inherent purposes, we do not tend to think that way today. When we do use language suggesting that objects have purposes—as in saying that a ball "wants" to roll down the hill—this is merely metaphorical language expressing the physical forces of nature. In a scientific context, we might refer to inanimate objects as falling, reacting, enlarging, or accelerating, but we do not consider them capable of *valuing* anything or being moved by intrinsic purposes. Our existence, on the other hand, is pervaded by considerations of value. We cannot conceive of our lives divorced from the notion that our relations with other individuals are good or bad for us, as are various outcomes and circumstances.

The more one reflects on the subject, the more evident it becomes that our experience is inseparable from the idea of value. Consider the myriad aspects of our mental lives that carry a positive or negative valence. Although people's experiences naturally vary a great deal, our stream of thoughts and feelings is commonly filled with things we desire (or wish to avoid), approve (or disapprove), find delightful (or repulsive), and assess highly or poorly. Our consciousness is brimming with sensations and states of mind that we evaluate positively or negatively. We generally cherish experiences of amusement, satisfaction, and elation. Other feelings, like anguish and despair, we view negatively, preferring fewer of them. We also have evaluative reactions and opinions of an ethical variety. We judge some acts, laws, institutions, and lives as admirable, noble, virtuous, beneficial, or morally proper, and others as corrupt, vile, harmful, unprincipled, or vicious.

The fact that we value things is also evident in our behavior. The idea of value is intimately connected with the motivations for our actions. We act in pursuit of ends. That is, we want things and direct our actions toward attaining them. The very idea of seeking a goal involves ascribing value to the end sought. Our lives would be unrecognizable without the assignment of value to the hopes and anticipated outcomes that underlie our actions.

We can appreciate the ubiquitous role of value in our lives from additional perspectives, including the manner in which we perceive and communicate about the world. Regarding the former, consider the staggering amount of information that a creature might potentially be equipped to collect about its environment. Given the complexity and diversity of the external world, it would be impossible for any living thing to process it whole. Any organism's biology inevitably reflects selections about the aspects of the environment that serve its survival and reproduction. This is true of human beings as well. We are equipped to take in only a sliver of the information that is potentially available. Our bodily senses have plainly been shaped by the kinds of knowledge that serve our survival and reproduction. Moreover, even though human biology enables us to perceive our surroundings to only a limited degree, no individual could fully exploit the entire array of sensory inputs with which we are bombarded. Through processes both conscious and subconscious, we make determinations about where to direct attention in navigating our environment, and these selections cannot help but be guided by value assessments.

If our acquisition of information about the world is guided by value, so too is the way that we form and exchange thoughts about the world:

communication through language. While the extent to which thinking might be possible without language is a thorny question, we clearly rely on verbal expressions to develop and share ideas. Value inescapably permeates this activity because developing a language entails numerous selections, which are inevitably influenced by the interests of those using the language. The most obvious way that language reflects choices is by according names to some things and not others. It is notable in this respect that we do not conceive of the world as an undifferentiated whole but as containing different kinds of objects. Given our dependence on the idea that reality comprises distinct things, we linguistically carve the world up into different classes of objects, and these choices too are value-laden. We treat some things as meriting notice, while other aggregations of atoms are never graced with a name.[3] When the need arises, we incorporate new terms into the language, while allowing words that no longer serve anyone's ends to fade from the scene. English has a word for the act of pouring water down one's throat but not on one's second toe, and if people anywhere have given a name to pouring water on the second toe, we can be sure that it has some kind of value for them.

Any proposed ultimate explanation must account for the existence of creatures for whom many things have value (or disvalue). If one grounds an account of existence in either science or mathematics, the notion that value has a place in the cosmos at all may seem mysterious. Value seems to have a fundamentally different character from the kinds of things that science and mathematics rely on to provide explanations. It is true that one may view the physical world as ultimately rooted in either science or mathematics while also regarding it with awe and reverence. Certainly, those endorsing scientific or mathematical approaches to the puzzle of existence may affirm value in the cosmos and our place in it. As we have seen, however, scientific and mathematical approaches to the puzzle of existence do not view the universe as sourced in value or as according a special significance to human beings. In seeking an objective basis for our beliefs, both approaches establish great distance between their explanatory bedrock and the distinctive features of human existence. That yawning gap presents a challenge for science and mathematics in furnishing full accounts of the manner in which we encounter the world. How could agglomerations of subatomic particles or fixed abstract relationships give rise to creatures like us: conscious beings who experience the world in all its richness while seeking meaning and contentment? There seems to be a difference in kind between human experience and the quantum fields of contemporary physics or the formal structures described by mathematics.

3 Leslie Armour, "Values, God, and the Problem about Why There Is Anything at All," *The Journal of Speculative Philosophy* 1, no. 2 (1987): 152–53.

In itself, of course, pointing to this challenge does not discredit scientific or mathematical approaches. It does, however, underline an advantage of an axiological approach to the puzzle of existence: namely, that it builds value into reality at the ground floor rather than having to account for its later emergence. By identifying goodness as the reason why the world exists, axiology avoids having to explain how the value-laden aspects of our lives could arise from something that is fundamentally value-free. An axiological universe is one in which the things that matter most to us have a similar fundamental character to the basis of the world's existence. From the perspective of an axiological theory, the pervasiveness of value does not present a problem. To the contrary, it manifests the elementary status that value enjoys in the cosmic order.

Teleological Arguments and the Fine-Tuning Problem

Teleological arguments are most familiar in the context of arguments supporting the claim that the world is rooted in the designs and intentions of an all-good divine person, commonly referred to as God. Nevertheless, they may also be used to support the claim that the world is based in goodness itself. In general terms, teleological arguments (based in the Greek term *telos*, meaning "end" or "purpose") aim to show that some outcome resulted from a process that was directed toward that end. In the context of discussions regarding cosmic origins, a teleological argument is meant to establish that certain good features of the world did not come about by chance, but, rather, as the product of purposive events.

We can think of teleological arguments as typically encompassing two main parts. The first part presents a puzzle by identifying features of the world that seem both extraordinarily good and surprising or unlikely. The second part offers a solution, contending that the best explanation for the existence of certain good things is that the world was directed toward them from the start. If successful, a teleological argument persuades you that there is a genuine enigma that would be difficult to address without an assumption of the world's inherent tendency toward the good.

For millennia, religion has been the principal context for discussion of the good as a reason why the world exists. Consequently, discussion of teleological arguments, too, has mostly taken place within a theistic framework. In that context, the good typically enters the picture as part of a deity's character, intentions, and actions. For example, "design arguments" aim to show that the best explanation for the existence of an essentially good world is that a good God created it. Nevertheless, since teleological arguments concern the world's directedness toward worthwhile ends, their basic structure is the same when used to show that the explanation for the world's existence is goodness itself rather than a good God. The shared idea is that certain good features of the world have resulted from an end-directedness built into the cosmos at the foundation. Axiological versions of the teleological argument substitute the good—standing alone—for God as the reason for the world's end-directedness.

Most often, the puzzle that serves as the jumping off point for a teleological argument focuses on the existence of life, and especially intelligent life. The mystery concerns the degree of coordinated complexity required to sustain advanced organisms. Human life depends on the interplay of numerous interlocking systems. Any one of them alone—say, the digestive, circulatory, or pulmonary systems—is remarkably complicated. But the coordination needed for all of them to work together boggles the mind. If life were simpler, we could more comfortably ascribe it to mere chance. However, a teleological argument insists that the emergence of life as we know it by mere coincidence strains credulity. The better explanation proposed is that the world was directed toward the good end of intelligent life.

Since teleological arguments have a similar basic structure when used to support either axiological or theistic conclusions, we can gain insights about them by considering versions that rose to prominence in a theistic context. A particularly influential theological version was advanced by William Paley (1743–1805) in his *Natural Theology: Or, Evidences of the Existence and Attributes of the Deity* (initially published in 1802). Although Paley presented his argument to support the existence of God—as understood by English Anglicans—his writing sheds light on teleological arguments more generally. His book presents a famous thought experiment. Imagine you are walking through an open field when you spot a watch at your feet. What would you conclude about how the watch had come to be? In Paley's view, no reasonable person would suppose that the watch had come into existence merely by chance. When one reflects on the multitude of improbable events that would have to take place for a watch to arise serendipitously, the argument goes, it becomes evident that coincidence is an implausible explanation for the watch's existence.

At the same time, you might have little difficulty believing that a small pebble at your feet was merely the product of chance occurrences. Paley's argument hinges on the difference between the watch and the pebble. Why is it less likely that the watch originated accidentally? For Paley, the reason was clear: only with respect to the watch do we observe that it has many parts arranged in a complicated manner to fulfill a particular purpose.[4] It

4 William Paley, *Natural Theology: Or, Evidence of the Existence and Attributes of the Deity, Collected from the Appearances of Nature* (Cambridge University Press, 2009), 1–4.

is difficult to believe that the intricate parts of a watch could have come together accidentally in just the right way to keep time. By similar reasoning, Paley contended, we may reasonably conclude that Earth's living creatures resulted from the intentional purposes of a good and powerful designer. Organisms, after all, manifest even greater complexity than a time-piece.

The intellectual context in which Paley offered his arguments made them especially resonant. The Scientific Revolution—along with the Age of Enlightenment that soon followed—led people to crave reasoned grounds for their theistic convictions. Over a century after Newton's transformative discoveries, many were no longer satisfied with accepting Church-approved texts as definitive grounds for knowledge about the natural world. Instead, inquisitive people increasingly sought justifications for their beliefs that were based in rational arguments informed by empirical observations. Theologians like Paley responded to that hunger by presenting a case for the existence of God that rested on epistemically respectable reasoning.

Although the teleological argument did not rely on experiments or anything that we would consider scientific evidence, it could nevertheless be seen as broadly empirical at least in the sense that its starting point was not Scripture or Church authority, but a set of observations. Gazing upon the world, one recognizes extraordinary complexity in the service of particular ends, as in, for instance, an eyeball's exquisite composition. Advances in scientific technology had intensified the sense of wonder regarding life's origins by enabling observers to peek behind nature's curtain as never before, revealing the biological world's complexity in greater detail. For example, improvements in microscopes brought to light a hitherto invisible world teeming with life, and highlighted the intricate designs and patterns inherent in living organisms.[5] For those who found Paley's teleological argument compelling, belief in God's existence could be framed as a reasonable response to puzzling observations, since it provided the best available explanation.

A second aspect of Paley's intellectual context that made his work especially resonant concerned beliefs about the relation between goodness and the natural world. For centuries prior to the Scientific Revolution, the dominant model of science was one attributed to Aristotle, whose ideas—for historical reasons related to the availability of his texts—exerted their greatest impact over a millennium after his death in 322 BCE.[6] In Aristotle's concep-

5 Catherine Wilson, *The Invisible World: Early Modern Philosophy and the Invention of the Microscope* (Princeton University Press, 1995), 85–90.

6 Having been lost for ages, Aristotelian thought became available in the West during the twelfth and thirteenth centuries when Arabic versions of Aristotle's texts were translated into Latin. While adaptations were needed to avoid conflicts with Church doctrines, prominent theologians, like St. Thomas Aquinas (1225–74), synthesized Aristotle's philosophy with Christian doctrines. The resulting dogmas became so established that they were called Scholasticism, reflecting their central role in the medieval educational program. Richard Tarnas, *The Passion of the Western Mind: Understanding the Ideas That Have Shaped Our World View* (Harmony Books, 1991), 175–78.

tion of the natural world, goodness and nature are inextricably intertwined. According to Aristotle's natural philosophy (or what today we would call "science"), individual objects of a certain kind share an essence—a form—that makes them the distinctive things that they are.[7] Crucially, the essential nature of a thing includes the good ends toward which it is directed. Thus, a central task for the natural philosopher is to study what constitutes the good for different kinds of things. For instance, an acorn's activities are aimed toward its actualization as a healthy, thriving oak tree. Given these premises, it made sense for Aristotle to frame the study of ethics as an investigation into the natural good for a human being.[8] Focusing on what it means for human beings to achieve their most characteristic function with excellence, Aristotle concluded that the highest good and end for human beings was flourishing or happiness over the course of a life.

The most salient point is that, for Aristotle, each thing's natural good has an objective reality in the world, because it represents the inherent good for that kind of thing.[9] However, the methods inaugurated by the Scientific Revolution rejected the snug fit between goodness and the study of nature that Aristotelian science took for granted. The new science viewed the world as playing out mathematically expressed laws that applied equally to all material things. Where the medieval synthesis of Aristotelianism and Christianity—known as Scholasticism—had *differentiated* objects in *qualitative* terms, modern science *unified* nature in *quantitative* terms. Observable changes in the natural world were now to be explained in terms of value-neutral physical regularities.

By rupturing the connection between nature and value, modern science intensified the mystery behind the existence of extraordinarily good things in the world. How could a natural world governed by indifferent, mechanistic, laws give rise to the marvelous complexity of intelligent life? Paley's teleological argument offered an answer suited to the intellectual climate, one that an educated person could embrace without embarrassment. The argument was based in reason—not Church authority—and it did not find purpose in the natural ends of particular things as in Aristotelian thought. Instead, purpose came in at the level of cosmic design. Confronted with the puzzle posed by the existence of intelligent life, the best explanation was that an all-good designer had purposefully created it. Modern science was valid, but in itself inadequate to account for the goodness evident in the creation.

---

7 Adamson, *Classical Philosophy*, 230–32.

8 See, for example, Aristotle, *Nicomachean Ethics*, trans. Joe Sachs (Focus Publishing, 2002), 1097a14–1097b7.

9 In this vein, the first sentence of Aristotle's *Nicomachean Ethics* proclaimed that "the good is that at which all things aim." 1094a 2-3.

Although Paley's teleological argument did not necessarily contradict prevailing scientific views, it was undermined half a century after his death by the work of a biologist, Charles Darwin (1809–82). Because teleological arguments present their conclusions as the best available explanation for observations about the world, they are vulnerable to the emergence of competing accounts. While Darwin was not motivated by antipathy to religion, his seminal work—*On the Origin of Species* (1859)—weakened the teleological argument by furnishing a plausible alternative explanation for the development of complex life forms.[10] Darwin's theory of natural selection showed how apparent design could arise without a designer.

The theory was based on the observations that offspring inherit many of their parents' traits and that only organisms surviving to the stage of reproduction pass on their traits to the next generation. The upshot is that traits conducive to survival increase in the population at the expense of those that are less advantageous.[11] Over many generations, the natural selection of the traits most favorable to survival generate increasingly complex organisms. Although Darwin could not explain exactly how parents transmit traits to their offspring, his theory assumed that this occurred through natural, impersonal processes.[12]

One might have expected that the theory of evolution would lead to the teleological argument's permanent decline. After all, the damage that Darwin's work inflicted on the teleological argument looks like a quintessential instance of science working against theology, and remarkable scientific discoveries continued in subsequent decades. In fact, however, developments in contemporary science—particularly in physics and cosmology—have made the teleological argument more pertinent than ever. Twentieth-century science breathed new life into teleological reasoning by rewriting the mystery of life in a form not resolved by Darwin. While natural selection explained the movement from simpler to more complex organisms, it did not explain how the first organisms came to be, as Darwin himself acknowledged. Evolutionary theory said nothing about why conditions

10 The book's original title was *On the Origin of Species by Means of Natural Selection, or the Preservation of Favoured Races in the Struggle for Life*.

11 Charles Darwin, *The Origin of Species* (P.F. Collier, 1909), 93–94.

12 In the absence of our modern knowledge of genetics, Darwin proposed a theory of heredity, "pangenesis," in which he hypothesized that all parts of an organism produce minute particles—gemmules—that collect in the reproductive organs and are passed on to the next generation. See Charles Darwin, *The Variation of Plants and Animals under Domestication* (John Murray, 1868). While Darwin's theory was highly innovative for its time, it was soon superseded by the discovery of the principles of genetics, particularly the work of Gregor Mendel on pea plants, which provided a more accurate picture of heredity. See Gregor Johann Mendel, "Experiments concerning Plant Hybrids," in *Proceedings of the Natural History Society of Brünn* IV (1865): 3–47. Ironically, we now understand that genetic variation involves random mutations, which means that the mechanisms producing the appearance of design in complex life forms depend on its very opposite: arbitrariness.

were favorable to the formation of life in the first place,[13] a question known today as the "fine-tuning problem." The scientific advances that brought this challenge to the fore were no more motivated by a desire to revive teleological arguments than Darwin's theory was aimed at undermining them. Nevertheless, the fine-tuning problem makes the challenge of finding plausible explanations for intelligent life more compelling than ever.

The fine-tuning problem focuses on features of the universe that were in place long before amino acids began cooking in a primordial soup or any fish walked onto the shore.[14] The problem gained prominence as scientific discoveries spotlighted specific features of the universe that are necessary for life to emerge. While the question remains controversial, many physicists believe that the odds were heavily stacked against the universe having any of numerous features that are indispensable to life. If the universe having even *one* such feature was extremely unlikely, then the prospect that it would possess *all* of them seems close to impossible. Yet here we are. What could be more evident than the fact of our own existence? Plainly, the universe does possess the features needed to support life. The mystery at the heart of the fine-tuning problem lies in the universe's particular characteristics. It is puzzling why the universe possesses all of the features necessary for life given the extremely low probability of that being the case.

Like the riddle that was the jumping off point for Paley's design argument, the fine-tuning problem centers on the improbability of complex living organisms. However, this version of the problem moves in directions unavailable to scientists in the eighteenth and nineteenth centuries. Paley had offered an impressionistic argument centered on an analogy meant to prime intuitions. By contrast, the fine-tuning problem is based in calculations derived from sophisticated models of contemporary cosmology and quantum physics.

The case that the fine-tuning problem identifies something that genuinely cries out for explanation is based on three critical assumptions about the nature of the universe. The first assumption is that each fundamental feature that the universe exhibits could have taken on a wide range of values. The universe's fundamental features include constants and behavioral regularities, such as the speed of light in a vacuum and the relative mass possessed by different subatomic particles. As noted earlier, contemporary physics has no explanation for why our universe has the particular natural laws and characteristics that it does.[15] Thus, this first assumption resonates

---

13 See J. Peretó, J.L. Bada, and A. Lazcano, "Charles Darwin and the Origin of Life," *Origins of Life and Evolution of Biospheres* 39, no. 5 (2009): 397.

14 Neil Manson, ed., *God and Design: The Teleological Argument and Modern Science* (Routledge, 2003), 2–4.

15 See Chapter Three.

with the predominant scientific view that there is no necessity underlying the specific set of behavioral regularities that the universe embodies. The second assumption holds that there was an equal likelihood that each fundamental feature would have taken on any one of its possible values. The third assumption states that only a tiny sliver of the possible values for each feature would have made life possible. Together, these assumptions suggest that there was an extremely low probability that any given fundamental feature would have taken on a value that allowed for the emergence of life.[16]

These assumptions would identify a puzzle even if they applied only to *one* feature needed to support the emergence of life. In fact, however, a large number of our universe's features have a value within the tiny range friendly to life. To appreciate why the combination of features needed to support life is so unlikely, we must briefly consider how probabilities work. The probability that two independent events will *both* occur is calculated by multiplying the probability for each of the two events measured separately.[17] For instance, suppose that there is a 50 per cent chance that a flipped coin will land showing the head side (and a 50 per cent chance that it will land showing the tail side). The chance that a coin flip will yield heads two times in a row is one in four (.5 times .5). The chance of three heads in a row drops much lower: to only one in eight (.5 times .5 times .5, or .125), and so on. This means that the probability of numerous very unlikely events all occurring is extraordinarily tiny. Imagine a fantastically gigantic roulette wheel with millions of pockets. In effect, the fine-tuning problem asks us to envision that the possibility of life in the universe depended on hitting the right single pocket not just once, but many times in a row. The odds against such a string of lucky spins are astronomically high. For those who view the universe's apparent fine-tuning as a true enigma, the challenge is to explain the occurrence of something that was so improbable.

A few examples will illustrate what physicists mean in speaking of the universe's finely-tuned features. Some of these features are easy to take for granted because they are so basic, but science offers no reasons to think they were inevitable. Most generally, the emergence of life requires a universe that is ordered. Consider the remarkable regularity in the ways that physical things behave. For example, as far as we can tell, an electron operates according to the same patterns of interaction with other physical things today as yesterday, and in our local precinct as in any other place in the universe.[18]

---

16 See Martin Rees, *Just Six Numbers: The Deep Forces That Shape the Universe* (Basic Books, 2000), 1–5.

17 Independence in this context means that the likelihood of each event occurring is not affected by whether the other event occurs.

18 Rodney Holder, *God, the Multiverse, and Everything* (Routledge, 2016), 6.

Without that kind of order and continuity, there would be no opportunity for the processes that produce life to play out over billions of years.[19]

More specific examples involve features of nature at the largest and smallest scales. The universe's rate of expansion sits on a knife's edge. If the expansion rate had been slightly faster, objects would have moved away from each other too quickly to allow for the formation of galaxies. With a slower expansion rate, the universe would have collapsed in on itself. In either case, the extraordinary complexity of living organisms could not have developed.[20]

The universe's fine-tuning also encompasses the relative values for the quantifiable characteristics of subatomic particles, such as mass and charge. Neutrons are roughly 0.1 per cent heavier than protons. If this differential was reversed—if neutrons were the lighter particle—protons would decay into neutrons, preventing the formation of atoms. Similarly, atomic nuclei could not be stable if the charge of electrons was even a tiny amount greater than it is.[21]

Other finely tuned features concern the four fundamental forces recognized by contemporary physics: the strong nuclear force, the weak nuclear force, electromagnetism, and gravitation.[22] The strong nuclear force (or "strong force") holds protons and neutrons together in atomic nuclei despite the mutual repulsion that the positively charged protons exert on one another. A slightly diminished strong force would be too weak to hold nuclei together. Yet if the strong force were slightly augmented, protons would bond together too easily to allow for the survival of hydrogen, whose nucleus has only one proton.[23] This scenario would also have made life impossible because hydrogen plays critical roles both as a component of water and as the primary fuel for stars.

---

19 Davies, *The Goldilocks Enigma*, 2.

20 Davies, *The Goldilocks Enigma*, 54, 149.

21 Lee Smolin, "Scientific Alternatives to the Anthropic Principle," in *Universe or Multiverse*, ed. Bernard Carr (Cambridge University Press, 2007), 328.

22 Physicists believe that the four forces are unified under conditions of sufficiently high energy, and they hope one day to gain an understanding that brings them under a single theory. Holder, *God, the Multiverse, and Everything*, 19; Greene, *The Fabric of the Cosmos*, 16. While progress has been made in unifying the weak and electromagnetic forces, the unification of the electroweak and strong forces remains elusive. Rees, *Before the Beginning*, 153. Moreover, the gap between gravitation and the other forces appears unbridgeable, presenting one of the most daunting challenges in contemporary science. See Chapter Three.

23 Martin Rees, *Just Six Numbers*, 232.

In itself, the fine-tuning problem is just that: a *problem*. There are many possible responses, and scientists disagree over how to address it.[24] The fine-tuning problem only becomes part of a teleological *argument* when one contends that the solution lies in recognizing the universe as directed toward good ends. The argument's persuasiveness depends on the relative strength of the competing explanations. One alternative is that the universe's fine-tuning resulted from chance. In that view, the universe's suitability for life is a brute fact with no explanation other than that coincidences do sometimes happen. The difficulty for the brute-fact response is the sheer scale of the coincidence, which makes it hard for many to take on board as a serious answer.

Another alternative posits that the cosmos is a multiverse encompassing countless universes that differ from one another. The key to the multiverse response is the assumption that each universe in the cosmic ensemble manifests a different combination of fundamental features. If there is an enormous number of universes with their own distinctive features, then there are vastly more chances that at least one will be friendly to life. When the fine-tuning problem asks how we happened to hit the winning lottery number, a multiverse theory replies that we did not buy only one ticket. How can you lose when you play every number? The multiverse response is particularly appealing to those who are intent on approaching the problem from a strictly scientific standpoint that refuses to takes refuge in either coincidence or anything that sounds supernatural.

Like earlier versions, teleological arguments based in the fine-tuning problem are most prominent in debates over the existence of God. Nevertheless, axiological theorists also use the fine-tuning problem to support the claim that goodness itself is the reason for the universe's friendliness to intelligent life.[25] The argument draws appeal from its reliance on cutting-edge research. It does not reject science, but recognizes a division of labor. Science

24 The fine-tuning problem's seriousness remains a topic of considerable controversy. For a number of reasons, some deny that the fine-tuning problem presents a genuine mystery at all. See, generally, Victor Stenger, *The Fallacy of Fine-Tuning: Why the Universe Is Not Designed for Us* (Prometheus Books, 2011). For example, some scientists and philosophers contend that life could potentially adapt to a wide range of different physical conditions. Those making this argument say that it is easy to erroneously assume that life requires our universe's precise conditions because that is all we know, but life might just as well be possible under a different set of laws and constants. In short, the claim is that the conditions for life are not so narrowly constrained as the fine-tuning problem suggests. Stenger, *The Fallacy of Fine-Tuning*, 291. Relatedly, others claim that our thinking is clouded by a selection bias; our universe's particular properties seem uniquely hospitable to life because it is the only one that we can observe. Simon Friederich, "Fine-Tuning," in *The Stanford Encyclopedia of Philosophy*, ed. Edward N. Zalta and Uri Nodelman (2023), § 3.2, accessed December 10, 2024, https://plato.stanford.edu/archives/win2023/entries/fine-tuning/. Yet another criticism charges that the fine-tuning problem's proponents are wrong to estimate the likelihood of the universe's suitability for life by varying only one constant at a time. Such an approach, the argument goes, overlooks the possibility that a change in one parameter might compensate for a simultaneous change in another parameter. Stenger, *The Fallacy of Fine-Tuning*, 280.

25 Below we discuss how John Leslie uses the fine-tuning problem to support his larger case for an axiological theory of existence.

elucidates the intricate ways in which physical things behave and interact, while an axiological perspective explains why the physical world takes a shape allowing for the emergence of creatures like ourselves.

Platonic Roots of Value-Based Theories

We have seen that contemporary axiological thought draws on the latest science in the form of the fine-tuning problem. At the same time, it also reflects the influence of philosophical thought tracing to Plato. Strictly speaking, Plato did not propose an axiological theory of existence in the sense that we are using that term here. That is, he did not suggest that goodness itself explained why anything physical existed. As discussed earlier, the ancients—including Plato—assumed that some kind of material stuff always existed, and they did not address why anything physical exists at all.[26] Nevertheless, Plato, like other ancient philosophers, was deeply interested in why the cosmos takes the shape that it does, including why it embodies such order and harmony. In this respect, Plato accorded a central role to goodness. Thus, before examining contemporary philosopher John Leslie's axiological theory in the next section, here we discuss the Platonic roots of ultimate explanations based in value.

Belief in the essential importance of the good played a significant role in ancient thought. Plato developed the groundbreaking idea that reality encompasses two fundamentally distinct kinds of being.[27] One realm comprises the ever-changing material particulars that we call physical objects. The other realm comprises the perfect, eternal, and immaterial forms, which are not only superior to the particulars, but are also responsible for making the particulars the kinds of things that they are. Although the ancients assumed that undifferentiated matter had always existed, Plato's dualistic view opened the door to later theorizing that physical existence had an explanation outside of itself.

Plato's ontological beliefs (about what exists) had implications for his epistemological beliefs (about how we acquire knowledge). In Plato's day, as in ours, some people believed that the only way to obtain knowledge was through the use of our bodily senses. Plato rejected that view, insisting that we can gain knowledge of the immaterial forms even though we cannot perceive them with our senses. To better understand the forms, Plato engaged in an exacting process of questioning called "dialectic."[28] In dialectical analysis,

26 See Chapter Two.

27 See Chapter Two.

28 Plato, *Republic*, trans. Desmond Lee (Penguin Books, 1974), 510d–511b.

one does not always follow a direct route from point A to B the way that one does in a deductive proof. Often, dialectic involves probing weaknesses in initial assumptions and adjusting proposals in a sustained effort to get closer to the truth. Plato learned dialectic from his mentor and hero, Socrates (469–399 BCE), who challenged people in the marketplace to provide definitions of virtues—such as courage, wisdom, or justice—and other things that were highly valued. When successful in provoking responses to his inquiries, Socrates prompted further deliberation through follow-up questions designed to probe inadequacies in his interlocutors' responses.

Plato's reliance on dialectic was based on his underlying conviction that the world is rationally ordered and that the most profound questions have answers that are largely open to our understanding.[29] Reality is neither haphazard nor beyond our comprehension. Our ability to grasp fundamental truths means that rational reflection on the way that we think and talk about the world may reveal insights into the cosmic order. Sensory impressions are not the only—or even the most significant—grounds for pursuing truth. While contemporary axiological thinkers do not necessarily employ dialectic in the same way that Plato did, they share his belief that sensory perception is not the only path to knowledge. We can gain better answers to fundamental questions through reasoned reflection.

Two Platonic dialogues—the *Republic* and the *Timaeus*—have been especially influential in the development of axiological thought. In the *Republic*, Socrates (the main character) explains that the perfect forms have a foundational relation to the good. As ideals, the forms encapsulate standards of excellence that particulars may reach toward but never fully realize. For instance, any particular thing in the world can only exhibit beauty to the extent that it imperfectly shares in the form of beauty. Even among the perfect forms, however, the good has a distinctive and superior role. The *Republic* proposes that the good is the basis for all knowledge and existence.[30] To illustrate, Socrates draws a parallel between the form of the good and the Sun, which not only provides the light that makes it possible for us to see but also generates the things that we see. Just as the Sun illuminates and gives life, the good illuminates the forms and is responsible for their existence.[31] Moreover, the good is also the object of our deepest love. It is something for which we rightly yearn.[32]

29 Tarnas, *The Passion of the Western Mind*, 32–33.

30 Scott MacDonald, *Being and Goodness: The Concept of the Good in Metaphysics and Philosophical Theology* (Cornell University Press, 1990), 4.

31 Plato, *Republic*, 504a–505c, 509b–511e.

32 This is a point made by a character named Diotima in Plato's *Symposium*. Diotima explains that one who loves wisdom must, through passionate (or erotic) effort, ascend through several steps from loving a particular beautiful body to ultimately loving the Form of the Beautiful and engendering actions of true virtue. Plato, *Symposium*, trans. C.J. Rowe (Aris & Phillips, 1998), 210a5–211d1.

Nevertheless, the Platonic good is not something that we can hope to fully grasp. Although the good is the source of being, it is fundamentally distinct from it. The good is beyond the reality that we can know and "superior to it in dignity and power."[33] When pressed for details by other characters, Socrates admits that his ability to provide answers is limited because the good remains difficult to wholly understand.[34] This does not mean that the good lies completely outside our understanding. Just as we can observe the things for which the Sun is responsible, so we can learn about the good from studying the things that derive their reality from it.

Plato's *Timaeus* built on themes from the *Republic*. The *Timaeus*' titular character claims that a rational designer—called the demiurge (or craftsman)—shaped the cosmos. Regardless of whether Plato wanted us to take the story literally—which remains uncertain—the dialogue offers an evocative way of thinking about the relation between the forms and sensible things. Equipped with a mind and the best intentions, the demiurge uses the perfect forms as a blueprint for crafting the cosmos.[35] Thus, the cosmos exemplifies beauty and goodness because a good and rational designer fashioned it on the basis of eternal ideals.[36] In explaining "how this created universe came to be made by its maker," Timaeus declares that the demiurge

> was good and ... being free of jealousy, he wanted everything to be as similar to himself as possible.... For the god wanted everything to be good, marred by as little imperfection as possible. He found everything visible in a state of turmoil, moving in a discordant and chaotic manner, so he led it from chaos to order, which he regarded as in all ways better.[37]

To be sure, Plato was not the first to speculate that a rational mind played a crucial role with respect to the order in the cosmos. Most notably, Anaxagoras (510–428 BCE) wrote of an infinite and self-ruled *nous* (the Greek term for mind or intellect), an overarching principle that imbued the cosmos with order and harmony.[38] However, Plato's thought was distinctive in finding the basis for cosmic order in transcendent ideals, and especially the form of the good. Since the forms ground objective truths about

33 Plato, *Republic*, 509b.

34 Plato, *Republic*, 506d–507a.

35 Andrew Gregory, "Introduction" to *Plato's Timaeus and Critias*, trans. Robin Waterfield (Oxford University Press, 2008), ix; Gabriela Roxana Carone, *Plato's Cosmology and Its Ethical Dimensions* (Cambridge University Press, 2005), 29.

36 Plato, *Timaeus*, trans. Robin Waterfield (Oxford University Press, 2008), 29a.

37 Plato, *Timaeus*, 29e–30a.

38 W.K.C. Guthrie, *A History of Greek Philosophy, Vol. 2: The Presocratic Tradition from Parmenides to Democritus* (Cambridge University Press, 1965), 293–97.

the best way for human beings to live, Plato's account merged cosmology with ethics in an unprecedented manner. Our aspiration to live well can be guided by manifestations of goodness in the cosmos itself.[39]

For those steeped in the Abrahamic religious traditions, the *Timaeus*' craftsman narrative cannot help but bring to mind the Creator God described in the Scriptures. But there are important differences. The biblical religions emphasize God's omnipotence, including the ability and intention to create the material world from nothing. By contrast, Plato's divine craftsman, driven by intrinsic goodness and the aspiration for excellence, looks to the forms as a blueprint for shaping the cosmos from pre-existing matter.[40]

Christian doctrine is one major conduit through which Platonic thought has had a tremendous reach. Early Christian theologians incorporated many Platonic ideas, which had largely been passed down to them through Neoplatonist thinkers, like Plotinus (204/5–270 CE). Several centuries after the great teacher's death, the Neoplatonists systematized Plato's dialogic explorations into a hierarchical framework that saw being as emanating from a single source known as the One, or simply the Good. When early Church Fathers began adapting pagan ideas to fit their Christian world view, many retained the Neoplatonist notion of a cosmic hierarchy. They placed an all-good source of being (the trinitarian God) at the pinnacle, followed in descending order by angels, human beings, and the beasts over whom people had dominion.[41]

In the last chapter, we saw that Tegmark's mathematical explanation for the world's existence incorporates Plato's view that mathematical objects exist eternally and independently of material things. However, in contrast with Tegmark, Plato did not view mathematics as the foundation of existence. Unlike Tegmark—who views everything in the cosmos as fundamentally mathematical—Plato's world view is deeply hierarchical. Plato recognizes a stratified relationship not only between the forms and particulars—with the former standing superior to the latter—but also among the forms. Mathematical objects are eternal, changeless, and perfect, but

---

39 Gregory, "Introduction," xv; Gail Fine, *Plato on Knowledge and Forms: Selected Essays* (Oxford University Press, 2003), 204–10.

40 Sarah Broadie, *Nature and Divinity in Plato's* Timaeus (Cambridge University Press, 2012), 11–12, 20; Francis MacDonald Cornford, *Plato's Cosmology: The* Timaeus *of Plato* (Hackett, 1997), 34–39.

41 Broadly speaking, some of the most influential early Christian thinkers—including Augustine of Hippo and Pseudo-Dionysius the Areopagite—adopted a Neoplatonist cosmic hierarchy. At the same time, there was a good deal of disagreement among Christian thinkers on the best way to integrate ancient and Christian thought. Peter Brown, *Augustine of Hippo: A Biography* (University of California Press, 1967), 503–05. Some early Christian theologians, like Tertullian and John Chrysostom, developed differing theological frameworks and cosmologies. Nevertheless, Augustine's Neoplatonism arguably established it as the dominant influence in shaping the theological and doctrinal landscape of early Christianity. Robert J. O'Connell, *St. Augustine's Early Theory of Man, A.D. 386–391* (Harvard University Press, 1968), 4–5, 13.

they occupy a lower rung on the cosmic ladder than other forms. Only one form reigns supreme: the good, which is the source of all being and understanding.[42]

For Plato, the ultimate aim of studying mathematics is to apprehend the good. Because mathematical objects are flawless, studying them makes it easier to recognize the good, which is the "region of ultimate blessedness" that we "must spurn no effort to see."[43] Nevertheless, mathematics cannot capture what is most essential to our lives, including the moral virtues crucial to the pursuit of a well-conducted life. Mathematics is quantitative. It involves precisely defined relationships, and axioms that are postulated up front. The good for a human being, though, is not like that. Pursuing the good requires the sustained discursive exploration of dialectic. Thus, for Plato, mathematical knowledge holds an intermediate status, superior to the particulars, but less exalted than the good.[44]

It is worth pausing to consider why our exploration intersects more with Plato's ideas than those of the other colossus of Greek thought, Aristotle. As Plato's most famous student, Aristotle agreed with his teacher that there are objective standards of goodness. However, an axiological theory entails not only that goodness is objective but also that it is independent of the material world. Unlike Plato, who championed both of these views, Aristotle embraced only the first. The pivotal difference between Plato and Aristotle concerns the forms. For Plato, the forms have an existence that is entirely independent of material entities. By contrast, Aristotle's forms are immanent in particular things; form and matter are intertwined, and the forms would not exist if there were no material things.[45] As a result, Aristotle's thought does not lend itself as readily to an explanation for the physical world that is based in goodness itself. Plato's notion of a transcendent good makes his thought especially attractive to later theorists seeking axiological explanations for the world's existence.

Leslie's Theory of "Ethical Requiredness"

Plato's influence shines brightly in the work of contemporary philosopher John Leslie (b. 1940). This section examines Leslie's defense of an axiological approach, and how he explains the world's existence as rooted in goodness.

42 Plato, *Republic*, 508c.

43 Plato, *Republic*, 526d.

44 Robert J. Baum, *Philosophy & Mathematics: From Plato to the Present* (Freeman, Cooper & Co., 1973), 16.

45 Aristotle, *Categories*, trans. J.L. Ackrill (Clarendon Press, 1963), 2b6 ("So if the primary substances did not exist it would be impossible for any of the other things to exist.").

Before delving into the details of Leslie's theory of "ethical requiredness," let us briefly consider its central elements.

Based in a Platonic assumption that the world is rationally ordered, Leslie assumes that the world's existence has an explanation. He then reasons that the basis of the world's existence must be non-physical (to avoid circularity) and must explain itself (to avoid an infinite regress). In fleshing out his theory, Leslie distinguishes between things that could *possibly exist*—although they do not—and things that *actually exist*. Leslie insists that there are truths regarding the relative goodness of possibilities for existence even in the absence of any actually existing things. Ultimately, Leslie finds grounds for the world's existence in the goodness of certain possibilities for existence: the fact that it would be good for certain things to exist is responsible for their actual existence. The theory avoids circularity because goodness is not itself a physical thing. It also avoids an infinite regress, because if we ask why goodness has creative power, the answer is because it is good that it do so. Goodness is explanation's final resting place.

The reason for examining Leslie's theory is not the extent to which others accept its claims. Indeed, many philosophers have attacked the theory. For some commentators, the sheer weirdness of the idea that goodness explains the world's existence is disqualifying. In this vein, scholars have called the theory "obscure,"[46] and "idiosyncratic, perhaps eccentric."[47] Others charge that Leslie does not adequately show how an *abstraction*—the requirement that good things come into being—could be responsible for the existence of *physical* objects.[48] An additional criticism is that theism better explains our own existence than Leslie's theory since God, like us, is an intelligent, personal being.[49] In this spirit, one scholar writes that it is "simply befuddling" to imagine standards of value apart from anyone for whom things could have value.[50]

Notwithstanding the criticisms, examining Leslie's theory will helpfully highlight distinctive and potentially appealing aspects of an axiological approach. Broadly speaking, the ascription of fundamental importance to goodness has long attracted followers, from Plato in ancient times to Paul Tillich and others in the twentieth century.[51] Although Leslie is not the only

46 Paul Helm, "Review of *Value and Existence* by John Leslie," *The Philosophical Quarterly* 30, no. 121 (1980): 376.

47 William J. Wainwright, "Review of *Value and Existence* by John Leslie," *The Philosophical Review* 90, no. 2 (1981): 320. See also John Mackie, *The Miracle of Theism* (Oxford University Press, 1982), 235; Robert Young, "Review of *Value and Existence* by John Leslie," *Religious Studies* 17, no. 1 (1981): 130.

48 Roland Puccetti, "Does the Universe Exist Because It Ought To? A Critique of Extreme Axiarchism," *Dialogue* 32, no. 4 (1993): 654; Del Ratzsch, "Review of *Value and Existence*, by John Leslie," *Noûs* 17, no. 1 (1983): 115–16.

49 Brian Hebblethwaite, "Review of *Universes*, and *Physical Cosmology and Philosophy*, by John Leslie," *Zygon: Journal of Religion and Science* 26, no. 2 (1991): 324.

50 Puccetti, "Does the Universe Exist Because It Ought To?," 654.

51 Mackie, *The Miracle of Theism*, 230–31.

prominent contemporary thinker who endorses an axiological approach to the puzzle of existence,[52] his theory is the most extensively developed. It is also the best known,[53] having garnered a great deal of attention from other scholars,[54] including those who praise Leslie's work as visionary,[55] "earnest, subtle and philosophically wide ranging,"[56] and "written with ... philosophical courage and much originality."[57]

Moreover, the criticisms themselves shed light on vital issues, such as whether any non-circular explanation for the world's existence must rely on abstractions. In addition, debates over whether the world can be better explained by divine creation or goodness itself date to ancient Greece and continue to the present day. Scholars also debate how much weight to place on whether a theory sounds unbelievable.[58] As we have seen, widely accepted theories in physics include claims that sharply contradict common sense.[59] When addressing the most profound and challenging questions, it is often the case that no candidate theories are free of counterintuitive implications. Regardless of one's ultimate assessment of Leslie's theory, then, there is much to be gained from considering it more closely.

To examine Leslie's theory in greater detail, we begin with his general defense of an axiological approach. Leslie offers two main arguments for recognizing the fundamental importance of goodness. The first argument builds from the Platonic assumption that the world has a rational order that is largely open to our understanding. Leslie begins by assuming that a rationally ordered world must have an explanation for its existence. The remainder of the argument reasons about the features that we should expect an explanation for the world's existence to have. According to Leslie, a satisfying explanation must rest on a foundation that is not physical in nature. After all, a physical explanation for the physical world would assume the very thing that it is supposed to explain. For this reason, Leslie argues, Edward Tryon's theory that the universe arose spontaneously in a

---

52 See, for example, Nicholas Rescher, *Axiogenesis: An Essay in Metaphysical Optimalism* (Lexington Books, 2010).

53 Puccetti, "Does the Universe Exist Because It Ought To?," 651.

54 See, for example, Hebblethwaite, "Review of *Universes*, and *Physical Cosmology and Philosophy*, by John Leslie," 317–24; Mackie, *The Miracle of Theism*, 230–39; John Churchill, "Paradox and Identity in *Theology* by B.T. Herbert, and *Value and Existence* by John Leslie (Review)," *The Thomist: A Speculative Quarterly Review* 45, no. 4 (1981): 630–39.

55 Helm, "Review of *Value and Existence* by John Leslie," 376–77.

56 Colin Lyas, "Review of *Value and Existence*, by John Leslie," *Philosophy* 55, no. 212 (1980): 276.

57 Hebblethwaite, "Review of *Universes*, and *Physical Cosmology and Philosophy*, by John Leslie," 324.

58 One philosopher observes that criticisms of Leslie's theory as unbelievable raise questions such as, "Given ... philosophical views which are [not] obviously incoherent, ... what makes some of them initially more plausible than others? Why are some views thought to be worth refuting and others not? Under what conditions are we entitled to dismiss a view as simply unbelievable?" Wainwright, "Review of *Value and Existence* by John Leslie," 321. See Chapter Nine.

59 See Chapter Three.

pre-existing quantum vacuum fails to explain why there is something rather than nothing. Far from being nothing at all, Tryon's hypothesized vacuum is itself an existing thing with physical properties.[60]

In addition to being non-physical, Leslie contends that an explanation for the world's existence must also be able to explain itself. Suppose E explains the world's existence. We will then want to know what explains E. And if A explains E, then we will want to know what explains A, and so on. This generates an infinite regress, which Leslie does not think can count as a satisfying explanation.[61] What, then, can stop the explanatory regress? Goodness, Leslie says, fits the bill. What explains why goodness is the explanation for the world's existence? The reason points back to itself: goodness has the feature of self-explanation because it is good that it does so.

Having appealed to the requirements of rational explanation, Leslie next presents a teleological argument centered on the fine-tuning problem. Leslie considers it remarkable that "events obey laws which satisfy life's very complex prerequisites."[62] The best response to this mystery, he argues, is to recognize that purposiveness is built into the fabric of the cosmos. A competing response to the fine-tuning problem is the idea of a multiverse. According to this alternative, the existence of countless universes with different characteristics makes it likely that at least one will be friendly to life. In rejecting the multiverse response, Leslie points to an overlooked aspect of the fine-tuning problem: life depends on the universe's fundamental features being set at the right value for more than one reason. That is, each feature must fulfill multiple interdependent requirements, which dramatically increases the complexity of arriving at a viable combination. As a result, there is no guarantee that even a countless array of universes would yield a single universe that could support life.[63] Leslie also rules out chance as an explanation for the universe's fine-tuning due to the extraordinary improbability that the universe would happen to end up with just the right features to support life. With the multiverse and chance eliminated, Leslie contends that the best explanation for the existence of intelligent life is the universe's inherent inclination toward goodness.

To develop a more fleshed out axiological theory of existence, Leslie builds on the Platonic belief that we can know many things beyond our sensory impressions. The landscape of truths encompasses not only logic and mathematics but also value. Although truths about value do not depend

60 John Leslie, *Immortality Defended* (Wiley-Blackwell, 2007), 73. Tryon's proposal is discussed in Chapter Three.

61 John Leslie, *Value and Existence* (Rowman and Littlefield, 1979), 4.

62 Leslie, *Value and Existence*, 6.

63 John Leslie, "A Proof of God's Reality," in *The Puzzle of Existence: Why Is There Something Rather Than Nothing?*, ed. Tyron Goldschmidt (Routledge, 2003), 136–37.

on physical existence, they are importantly germane to physical things. In describing how truths about value relate to physical existence, Leslie draws a distinction between possibilities for existence and actual existence. There are some kinds of things that cannot possibly exist because their existence would entail a logical contradiction (such as a round square). A possibility for existence is something that does not actually exist but that hypothetically could exist. For instance, although unicorns do not actually exist, it seems possible, given alternative circumstances, that horses with a horn on their foreheads *could* exist. It is crucial to Leslie's account that there are truths about *possibilities for existence* (truths about what could exist) even in the absence of any *actually existing things*.

Two kinds of truths about possibilities for existence figure importantly in Leslie's theory: logical and axiological. Logical truths rule out the existence of anything that entails a contradiction. Truths of this kind leave an extremely large number of possibilities for existence still in play, because there are so many possibly existing things that do not involve any contradictions. Unlike *logical* truths, *axiological* truths involve comparative evaluations concerning the goodness or badness of possibilities for existence. Even if X and Y are mere possibilities for existence, there are axiological truths about whether it would be better for X or Y to actually exist.[64]

Leslie uses thought experiments to bolster the plausibility of his claim that truths about possibilities for existence do not depend on actually existing things. He asks us to compare two possible states of affairs: one in which nothing exists but a group of people experiencing uninterrupted pleasure, and another in which the same people experience nothing but misery. Leslie asserts that the actual existence of the former would obviously be better than the latter.[65] He also envisions the case of a young person who dies in a tragic accident. Even though after the accident the person no longer exists, it is still true that it would have been better had the person survived. The crucial point is that there are objective axiological truths about people or states of affairs that no longer exist.

On an immensely larger scale, Leslie imagines that everything in the world is suddenly annihilated. It is an objective truth, he says, that this event was very bad even though nothing exists any longer. If a value claim can be true about things that *no longer exist*, he argues, then a value claim can also be true about something that *does not yet exist*.[66] In this vein, Leslie claims there are hypothetical states of affairs so awful that they are worse than absolute nothingness. Most significantly, there are also hypothetical states

64 Leslie, *Value and Existence*, 2.

65 Leslie, *Immortality Defended*, 2.

66 Leslie, *Value and Existence*, 2, 46.

of affairs that are better than nothingness. Even before anything existed, for example, it was an objective truth that the emergence of intelligent life would be "genuinely fortunate, truly good."[67]

The pivotal assertion in Leslie's theory is that the comparative goodness of some possibilities for physical existence is responsible for their actual existence. He refers to this relation between truths about value and existence as "synthetic necessity." This kind of necessity is not the same as logical necessity, because logical principles only rule out contradictions, which still allow innumerable possibilities for existence.[68] Neither is synthetic necessity the same as the kind of causal necessity that is familiar in scientific explanations. As noted, Leslie denies that science can explain the world's existence because it assumes the existence of physical things from the start. In further elaborating the idea of synthetic necessity, he notes that it is not temporal in nature. As Leslie puts the point, "That some factor 'created' the world need not mean that initially no world existed and then the factor suddenly became active."[69]

So, what *is* this elusive relation between value and existence that Leslie refers to as synthetic necessity? He describes it as the link between a particular kind of truth—that it would be good if something existed—and the fact that it actually exists. In Leslie's terminology, if it would be good for a thing to exist, then its existence is "ethically required."[70] ("Ethical" in this context is essentially a synonym for axiological, referring to value in the broadest sense; it does not suggest a specifically moral duty).[71] Put another way, ethical requiredness describes an internal relation between a thing's features (the qualities that would make it good for it to exist) and its own justification for being (the reason that it actually exists.)[72]

To appreciate what Leslie means by an "internal relation," consider that when two objects have an external relation, that relation can be altered without making changes to the objects themselves. For example, two objects connected to one another by a rope have only an external relation: that of being connected by the rope. The connection between the objects could be eliminated without altering the objects simply by cutting the rope. By contrast, an internal relation expresses something about an object's inherent features, and, thus, does not depend on anything outside itself. Leslie is insistent that ethical

---

67 Leslie, *Immortality Defended*, 24.

68 Leslie, *Immortality Defended*, 29–31; Leslie, *Value and Existence*, 3.

69 John Leslie, *Infinite Minds: A Philosophical Cosmology* (Clarendon Press, 2001), 174–75.

70 Leslie, *Value and Existence*, 36–38.

71 Leslie, *Immortality Defended*, 1–2.

72 Interestingly, there seems to be a remote precedent of this principle in Aristotle's philosophy as well, as he claimed that being is better than non-being. Aristotle, *Generation of Animals*, trans. A.L. Peck (Harvard University Press, 1942), 731b18–732a2.

requiredness—the relation between goodness and existence—describes only an internal relation. It does not involve something outside of a thing that could be added to it or removed. It is not as though existence comes first and then the property of goodness is slathered on top like a coat of paint. Rather, it is the case that the object shall exist due to the (good) kind of thing that it is.[73]

Thus, the synthetic necessity that characterizes the relation between value and existence is an authentic kind of necessity even though it does not involve logical or causal necessity. The term "necessity" here connotes the idea that ethical requiredness is not merely one consideration among many. In Leslie's terms, it "has a creative success that is necessary in an absolute fashion,"[74] and it "is something fulfilled, satisfied, when the required entity, activity, or state of affairs has actually come to exist."[75]

Leslie suggests that Plato may have had something like ethical requiredness in mind when he had Socrates proclaim that the form of the good was responsible for the world.[76] Socrates' indication that the form of the good was beyond existence resonates with Leslie's claim that even if nothing existed, there would still be an ethical call for it to exist.[77] Regarding the efficacious power of possibilities, Leslie writes that "the character which a situation *would have* can provide an ethical ground for it to come to exist."[78] It is only fitting that Leslie draws so much inspiration from the Platonic tradition, which locates the highest form of rational order not in math or physics, but, rather, in goodness itself.[79]

What Kind of Thing Is Value?

The proposal that value or goodness is the basis of physical existence raises questions about the nature of value, and, in particular, about the kind of existence that it has. After all, an explanation for the physical world that

73 Leslie, *Value and Existence*, 30–34.

74 Leslie, *Immortality Defended*, 87.

75 Leslie, *Immortality Defended*, 23.

76 Leslie, *Immortality Defended*, 2.

77 John Leslie, "A Cosmos Existing through Ethical Necessity," *Philo* 12, no. 2 (2009): 176–77.

78 Leslie, *Value and Existence*, 2.

79 Another prominent contemporary axiological thinker, Nicholas Rescher, also rests his framework on broadly Platonic assumptions. Nicholas Rescher, *Axiogenesis: An Essay in Metaphysical Optimalism* (Lexington Books, 2010); Nicholas Rescher, *The Riddle of Existence: An Essay in Idealistic Metaphysics* (University Press of America, 1984). Moreover, like Leslie, Rescher also relies heavily on the notion of possibilities for existence. At the same time, Rescher criticizes Leslie for employing a conception of value that is too centered on what is good for human beings in particular. He also criticizes Leslie's description of the relation between value and existence. As noted, Leslie claims that value bears a direct relation to existence as an internal relation between a thing's properties and its justification for being. By contrast, Rescher proposes the existence of "protolaws," which mediate the relation between value and existence by effectively governing what shall and shall not exist.

depended on physical existence would be circular, and, thus, unavailing. The previous chapter similarly examined the debate over whether mathematical objects exist independently of the physical world. In philosophical terms, the question we consider now concerns the ontological status of value.

The nature of claims about value fascinated the ancients and continues to captivate people today. Many people wonder at one point or another about whether there are objectively correct answers to questions about value, such as whether some ways of spending one's time are inherently more meaningful than others, or whether some artistic works are better than others. Indeed, the objectivity of value is probably among the most widely known philosophical issues (especially when it comes to questions about morality and justice).

Two principal questions about the nature of value bear on axiological theories: are there objective truths about value, and, if so, do these truths depend on the existence of physical things? Philosophers and others vigorously debate the subject. Although not everyone engaged in these controversies is specifically interested in why there is a physical world, their arguments nonetheless have crucial implications for our exploration.

Only one general position regarding the status of goodness is friendly to axiological theories of existence: "non-natural realism." This position combines two crucial claims about value: that it is objective (realism) and that it is independent of the physical world (*non-natural* realism). According to non-natural realism, there are propositions about value that are true apart from what any particular people believe, and the basis of these truths does not depend on anything that is physical in nature. The latter claim means that truths about value have a different character from truths about physical properties, such as the mass of a quark, or an electron's charge.

We begin with the debate over realism. Realists hold that at least some value claims are true regardless of what anyone thinks about them. For realism to be correct, it must be the case that (1) when people make statements about value, they are attempting to say things that *could be assessed as true or false*, and (2) at least some of the statements that people make about value *actually are true*. Opposition to realism takes many forms. One anti-realist position—"non-cognitivism"—rejects the first proposition, thus denying that the things people say about value even count as attempts to express propositions that could be true or false.[80] If people making statements about value are not attempting to say things that are true, then what on Earth are they doing? One influential view—"emotivism"—holds that statements involving value are really nothing more than exclamations conveying the

80 David Enoch, *Taking Morality Seriously* (Oxford University Press, 2013), 3.

speaker's feelings. An assertion that an act was immoral amounts to blurting out the interjection "Boo!" in its direction.[81] Another non-cognitivist position—"prescriptivism"—holds that remarks involving value are really imperative statements aimed toward others' behavior. In proclaiming that X is wrong, a person simply tells others not to do X.[82] Whatever its specific form, non-cognitivism denies that we use language to make truth-evaluable claims. Non-cognitivism does not even allow statements about value the minimal honor of being wrong.

A rival anti-realist position—"error theory"—agrees with realists that people make statements about value in an attempt to say things that are true. However, error theorists depart from realists in denying that people actually ever succeed in saying anything that is true about value. It is important to note that error theory is not merely a position about our current ability to get things right. Instead, it points to a basic reason why no statements about ethics ever could be true: because nothing actually has the property of being ethical (or unethical).[83] By analogy, no matter how many times people try to make true statements about werewolf physiology, they will always fail, because there is in reality no such thing as werewolves or werewolf physiology. Similarly, if there are no facts about what is ethical or unethical, it is impossible to say true things about what is ethical or unethical. (Like a good portion of the scholarly debate, much of the present discussion refers to ethical claims in particular, but the arguments discussed here nevertheless apply to value claims more broadly.)

Why do some deny that there are objective truths about ethics? One argument of this kind focuses on the easily observable fact that people often disagree on ethical questions. Now, in itself, the mere fact that people disagree about ethics is not a strong argument against realism, because it seems wrong to suppose that a consensus of opinion is a prerequisite for truth. We do not generally think that the truth of whether the Earth revolves around the Sun depends on the results of a public opinion poll. It was true that the Earth revolved around the Sun even in historical periods before most people recognized this fact. However, a more refined anti-realist argument observes that realism does not fit well with the way that ethical disagreements develop over time. For instance, if there were objective truths about ethics, we might expect that the extent of disagreement would decrease over time, converging around propositions shown to be the most reliable. We might also expect people with expertise on ethical topics to arrive at widely accepted positions. According to error theorists, there is a ready explanation

81 A.J. Ayer, *Language, Truth, and Logic* (Dover Publications, 1952), 67.

82 R.M. Hare, *The Language of Morals* (Oxford University Press, 1972), 1–2.

83 J.L. Mackie, *Ethics: Inventing Right and Wrong* (Penguin Books, 1990), 33.

for why people do not reach a greater degree of agreement on ethical truths: because there are no such truths.

Shifting to arguments in favor of realism, a significant motivation for those who embrace the position is the determination to make sense of the way that we think and talk about ethics.[84] To many people, few things seem more unshakable than the truth of certain propositions about right and wrong, such as the depravity of killing innocent people for no reason. Furthermore, we tend to associate claims about ethics with universal application. My belief in the immorality of purposeless killing is not like my preference for rye bread over pumpernickel. It comes easily to say that rye is a better choice *for me* while acknowledging that the same might not be true *for others*. Yet my convictions about purposeless killing are not like that. I cannot help treating the wrongness of murder as a proposition that is true for other people as well. For realists, the reason it seems so clear that certain statements about ethics are objectively true is that they are.

Consider as well the way that we debate value questions. When it comes to strongly held ethical views, most people do not merely take note of differences in personal taste and leave it at that. Instead, they offer reasons in an attempt to persuade other people to adopt their point of view. This widespread practice of engaging in arguments about ethical positions suggests an underlying belief that there really are true and false answers to ethical questions. An advantage that realism enjoys is that it does not need to explain away evident features of the way that we think and talk about ethics.

When they turn to addressing arguments made by anti-realists based in the way that people disagree about ethics, realists note that it is common to overestimate the actual extent of ethical disagreement. On closer inspection, realists contend, we can recognize that many apparently ethical disagreements in fact turn on non-ethical disputes.[85] Two people with absolutely identical beliefs about ethics might nevertheless have opposing opinions about capital punishment because they hold different views about its effectiveness as a deterrent. Moreover, there is considerable consensus around the globe on many significant ethical matters, such as the value of friendship or the general obligations that parents have regarding the care of their children. While it is undeniably true that some people transgress even the most widely held ethical propositions, that fact poses no difficulty for realism, because realists do not assert that everyone understands ethical truths, much less that they always follow them.[86] Ethical truths are complicated,

84 See, for example, W.D. Ross, *The Right and the Good* (Clarendon Press, 1930), 85, 89; Enoch, "Taking Morality Seriously," 1.

85 Philip Stratton-Lake, "Introduction" to W.D. Ross, *The Right and the Good* (Clarendon Press, 1930), xliv–xlv.

86 Stratton-Lake, "Introduction," xlvi.

and self-interest and other factors may get in the way of people acting ethically in practice. Nevertheless, none of these observations is at odds with the claim that ethical facts are objectively true.

As noted, an axiological theory of existence requires not only realism, but *non-natural realism*. For truths about value to explain physical existence, they must be not only objectively true but also independent of the physical world. Among realists, there is a divide of opinion about whether truths about value depend on the physical world. *Natural realists* maintain that ethical truths are ultimately based in physical things. For instance, some natural realists claim that ethical truths are rooted in the pleasures and pains that we experience as biological creatures, and others ground ethics in the satisfaction of desires. Natural realists claim that pleasures and pains, and the desire of satisfactions, are appropriately studied through the methods of natural science.

Natural realism has appeal for many because it seems less mystical than non-natural realism. A true proposition is true *about* something. That is, there must be something that makes it the case. What, then, makes ethical claims true? Natural realists say that things in the physical world make ethical claims true. For many natural realists, the notion that anything other than physical properties could make things true is just too weird or mysterious to take seriously.[87] In this vein, natural realists point to what they consider a bizarre implication of non-natural realism. If non-naturalism is true, they argue, then two distinct worlds could be identical in every physical respect—they could have the very same *physical truths*—yet nevertheless have different *ethical truths*. There is nothing in non-natural realism that would rule out this possibility. Yet natural realists do not see how anything could plausibly make the two worlds different from an ethical standpoint once it is stipulated that they are physically indistinguishable.[88]

Another motivation for many natural realists is that the position seems to have less difficulty than non-natural realism in explaining how we acquire ethical knowledge. Acquiring knowledge about something seems to require interacting with it in some way. You may discover that a mattress is firm by sitting on it, for example, or that something is baking in the kitchen by smelling the cinnamon rolls. If natural realism is correct, then it is easy to explain how we are able to learn ethical truths, since they are based in physical things, which we can interact with through our bodily senses. However, if ethical truths are not based in anything physical, natural realists ask, what kind of interaction could we possibly have with them? Denying any physical basis for ethical truths, natural realists argue, seems to require positing that we are endowed with some kind of exotic, extrasensory power.

87 See Mackie, *Ethics*, 38.

88 See Enoch, *Taking Morality Seriously*, 134–37.

In response, non-natural realists insist that values and physical things are too disparate for one to be based in the other.[89] Consider the difference between truths about gravity and truths about the obligation to keep promises. They seem to involve radically different kinds of properties. Non-naturalists deny the plausibility of viewing ethical claims as just another subcategory of scientific truths alongside physics and chemistry.

Another prominent line of reasoning that supports non-realism is known as the "open question argument." Developed by philosopher G.E. Moore (1873–1958) in the early twentieth century, the argument maintains that ethical properties—such as "goodness"—cannot be equated to natural properties. For any proposed identification of an ethical property with a natural one, the legitimacy of this identification always remains an open question. That is, for any physical property which natural realists claim is equivalent to an ethical property, it will be meaningful to ask in reply whether that physical property really is good. Suppose a natural realist claims that "goodness" is synonymous with "pleasure." It nevertheless remains legitimate to ask whether it is really true that what is pleasurable is also indeed good. Moore contends that the answer is not always self-evident or obvious. If the meaning of "good" and "pleasure" were equivalent, then inquiring about the goodness of pleasure would be redundant, like asking whether one equals one. Thus, Moore's argument concludes that ethical terms like "good" cannot be simply reducible to physical terms.[90]

Non-natural realists maintain, then, that ethics is a domain of objective truths, independent of physical properties, and governed by its own norms and principles.[91] Although non-natural realists have not all agreed on a single account of how we learn about ethics, many claim that ethical truths are self-evident, or open to our understanding through reflective reasoning. Drawing an analogy with the way that our bodily senses enable us to perceive physical things, others maintain that we directly apprehend values through a non-sensory faculty.[92]

The Specification Problem

In this section and the next we discuss objections to axiological theories. Before turning to the problem of evil in the next section, here we discuss an objection with a very different character. While the problem of evil charges

89 Enoch, "Taking Morality Seriously," 4.
90 G.E. Moore (1903), *Principia Ethica* (Cambridge University Press, 1956), § 13. See also Fred Feldman, "The Open Question Argument: What It Isn't; and What It Is," *Philosophical Issues* 15 (2005): 23–24, 35.
91 Stratton-Lake, "Introduction," xlii.
92 Stratton-Lake, "Introduction," xiii, xliv.

that the world does not look like one based in goodness, the specification problem questions whether there is enough content in objective truths about goodness to pick out a world for existence in the first place.

Let us recall what an axiological explanation of the world's existence entails. In an axiological theory, goodness is much more than merely an aspirational standard guiding our conduct. It must bridge the gap between the possible and the actual. From all of the possible worlds, goodness selects one for existence. Leslie's theory of "ethical requiredness" illustrates the role that goodness plays in an axiological theory. Leslie asserts that there are objective truths about the relative goodness of different things that hypothetically could exist. It is objectively true that it would be better if some things existed than others, and these truths bridge the gap between mere possibility and actual existence.

Is it plausible that there really are objective truths about the relative goodness of different possible existing things or states of affairs? Leslie offers various examples to support his claim that there are. A world with nothing but happy people would be better than one with only miserable people. It would have been better if a particular person had not died in a tragic accident. If the world were suddenly annihilated, it would be true that it would have been better had this not occurred. Similarly, a world with intelligent creatures like ourselves is better than one without them. It is hard not to notice that Leslie's examples sweep in extremely broad strokes. In this respect, they seem designed to be as obvious and uncontroversial as possible. In each case, the polar or extreme nature of the contrasts lends greater plausibility to the claim that one state of affairs must be objectively better than another. It may seem hard to deny that maximal happiness is better than unmitigated suffering.

However, even accepting that extremely broad claims of the kind in Leslie's examples are objectively true, an axiological theory demands much more than that from goodness. After all, the world does not only consist of things that are describable in sweeping contrasts, like happiness versus misery, or life versus death. The world is also teeming with green and purple flowers, guppies and highlighter pens, igneous rocks and vanilla beans. The specification of a particular world seems to require innumerable selections among a vast array of possibilities. Moreover, the world includes many details that are value neutral. It is difficult to imagine that it would be objectively better if Neptune were a little bigger, or there were slightly fewer grains of sand on a particular beach and a few more on another.

The specification problem questions whether an axiological theory has the internal resources to explain how goodness can pick out a particular world for existence. At the least, we can say that it would strengthen an axiological account to say more about the role that goodness plays in specifying a world. For instance, to what extent do non-axiological considerations limit the pool of possible worlds? How specific must truths about the relative goodness of possible existents be to fulfill the role of picking out a world? And how does goodness figure in the rendering of ethically indifferent choices?

Reflection on the specification problem brings out an intriguing contrast between axiological and mathematical theories of existence. As noted in the previous chapter, the role that human beings play in selecting axioms for a particular mathematical system is arguably problematic for a mathematical answer to the puzzle of existence. Any arbitrariness in selecting axioms might raise doubts about the necessity of the truths derived from them. Nevertheless, once the axioms for a mathematical system are in place, it is clear that we can derive an extraordinarily rich set of propositions from them. Thus, while there are countless mathematical truths, we cannot say that they are simply true (that is, invariably, for any mathematical system), but only that they are true within a specified system in light of its axioms. There is, then, no single set of axioms for all of mathematics, but, rather, many truths that can be demonstrated within a system given its unique set of axioms.[93]

In an interesting sense, the challenge facing an axiological approach is the opposite side of the coin to the one facing a mathematical approach. It seems plausible to suppose that there are certain foundational axiological truths that could not be otherwise (such as the preferability of happiness over misery). If so, there is a single set of axiological truths that serve universally and equally well for all reasoning about value and goodness. These axiological propositions are simply true rather than being true only given a particular set of assumptions. However, even if an axiological approach has access to such a universal set of foundational truths, it is not clear how much can be derived from them, especially if it turns out that the only reliable, general axiological truths are of an extremely general character. How many conclusions can we derive from the premise that unlimited happiness is preferable to unyielding misery?

The potential difficulty surfaces when we try to imagine how such broad propositions could generate the kind of detailed and rich content that a theory like Leslie's needs for goodness to pick out a world. The challenge that the specification poses for axiological theories is to show that truths about goodness do in fact have a sufficiently fine-grained texture to pick out one physical reality from the multitudinous possibilities.

---

93 As also noted in the previous chapter, many contemporary mathematicians do not believe that axiom selection poses any threat to the integrity or necessity of mathematics as a whole, since there is necessity underlying which collections of axioms can be used in a mathematical system.

The Problem of Evil

Any theory rooting existence in goodness must grapple with the apparent existence of bad things, a challenge that has long been known as the "problem of evil." Claims that existence is rooted in goodness cannot help but make us think about things in the world that seem anything but good. One scholar notes that it "would appear to be a consequence of any account that postulates a necessary connection between being and goodness—that everything that exists (or has being) is good,"[94] and yet that hardly seems to be the case. As discussed further in the next chapter, theistic explanations of the world's existence also face the problem of evil, because they commonly explain the world's existence as the creation of an all-good God. In fact, the subject attracts so much attention that it has generated an entire branch of theology— known as "theodicy"—dedicated specifically to that topic.[95] Nevertheless, while the discourse regarding the problem of evil has been developed most fully in the theological context, many of the salient arguments apply with equal force to axiological theories, which claim that goodness itself explains the world's existence.

We saw in the last section that the specification problem challenges the internal logic and resources characteristic of axiological theories, questioning whether goodness has sufficiently rich content to select a particular world from all the possibilities. By contrast, objections based in the problem of evil begin with observations about the nature of the world in which we find ourselves. The problem, in a nutshell, is that the actual world just does not look like one that originated in goodness.[96] Our reality is at odds with what we would expect to see if the claims of an axiological theory were true. The objection from evil contends that the world would be a more desirable home if goodness really was the base of its existence.

Arguments appealing to the problem of evil exert a strong, visceral pull. No one is immune from pain, disappointment, loss, sadness, and heartache. Even those who have been the most fortunate in their own lives are aware of the horrors inflicted on others, including those associated with war, famine, poverty, disease, slavery, oppression, and natural disasters.

---

94 MacDonald, *Being and Goodness*, 6.

95 Given the immense literature on the problem of evil, which has been intensely debated for millennia, we can only scratch the surface of the subject here by discussing a few of the more prominent arguments in the discourse. For examples of fuller explorations of the subject, see David Hume, *Dialogues concerning Natural Religion*, ed. Norman Kemp Smith (Thomas Nelson and Sons, 1947); John L. Mackie, "Evil and Omnipotence," *Mind* 64 (1955): 200–12; John Hick, *Evil and the God of Love* (Harper & Row, 1977); and Alvin Plantinga, *God, Freedom, and Evil* (Harper & Row, 1974).

96 T.J. Mawson, *Belief in God: An Introduction to the Philosophy of Religion* (Clarendon Press, 2005), 198–99.

The objection from evil begins from a similar place as the teleological argument—general observations about the nature of the world—but the former turns the latter on its head. The teleological argument highlights the existence of glorious things about the world, like its hospitability to life. Noting the extraordinary unlikelihood of a universe able to generate creatures like ourselves, the teleological argument contends that the cosmos is inherently directed toward good ends. Against the grain of this reasoning, the objection from evil stresses that an axiological account is hard to square with a full and candid account of what actually exists. Emphasizing things in the world that are far from good, it asks whether we would really expect a world built on a foundation of goodness to contain so many bad things, such as undeserved and unrelenting suffering.[97]

No matter where one's assessment of axiological theories settles in the end, few would deny that it *seems* like the world contains bad things. Consequently, an axiological theory owes us an explanation for why so many things appear to be bad within a world supposedly directed toward the good. A prominent strategy for furnishing an explanation of this kind contends that people fail to appreciate the true extent of the world's goodness. One purported reason why people assess the world's overall goodness in an excessively harsh manner is that they measure it against an impossible standard. More specifically, many incorrectly assume that all conceivable good things could exist simultaneously. Finding that reality does not match this unrealistic fantasy, they conclude that the world is not intrinsically directed toward the good. According to this defense of axiological theories, different kinds of goods may conflict with one another.[98] In this vein, Leslie says that there is a complex structure to the way that various goods interrelate. Highly ranked goods may "overrule" other goods with which they are not compatible.[99] The possibility for goods to conflict is supposed to explain why we do not find an unremitting appearance of good things in our lives.[100]

The classic example of a great good that is said to conflict with—and, thus, rule out—other goods is free will. While there is no agreed-upon definition of free will, those relying on free will to rebut the objection from evil typically have in mind individuals' ability to exercise agency over their choices (rather than having their actions fully determined or constrained by forces outside their control).[101] The free will argument assumes that human

97 Brian Davies, *An Introduction to the Philosophy of Religion* (Oxford University Press, 2021), 244–46.
98 Mawson, *Belief in God*, 203–05.
99 Leslie, *Value and Existence*, 7.
100 Leslie, *Immortality Defended*, 26.
101 Alvin Plantinga, *The Nature of Necessity* (Clarendon Press, 1974), 165–66.

beings actually have free will (a hotly contested philosophical issue), and that it is extremely good that they do. The power to make choices for ourselves is a magnificent thing in itself and makes possible other great goods, such as triumphs of personal effort, moral acts that overcome temptation, deserved pride in one's achievements, expressive creativity, and loving relationships developed through chosen commitments to others.[102]

Why does free will conflict with other goods? By its nature, free will makes it possible for people to make choices that are ill-considered, reckless, foolish, or intentionally immoral (evil). Endowed with free will, some people exercise it to triumph over adversity through honest effort, while others cheat or maliciously undermine their rivals. Some people exercise their free will to engage in morally admirable actions, while others do unjust and terrible things. Iniquitous decisions are bad in themselves and often lead to bad consequences: hurt feelings, disappointments, injury, pain, and suffering. We understandably yearn for a world where people only exercise free will in the best possible manner. But that notion is at odds with the nature of free will. If the world were arranged so that people could only do the best things, then they would not actually be acting freely. Thus, the free will response to the objection of evil crucially insists that the existence of free will cannot be separated from its downstream outcomes.[103] A world with only the patently good aspects of free will is impossible. Free will is a great good that inevitably brings with it the occurrence of things that, taken in isolation, seem far from optimal.

Even if the free will response is persuasive, the scope of its explanatory effectiveness is limited given that the world contains undesirable things that do not result from free choices. Long before there were people—or any other organisms for that matter—volcanoes erupted, torrential downpours brought floods, and boulders tumbled down mountainsides. Events of this kind continue to occur and cause tremendous destruction, injury, and suffering to millions of people. The free will response cannot fully explain away this sort of harm because it cannot be fully ascribed to the effects of freely chosen actions.[104]

To address harms resulting from natural disasters, defenders of axiological theories point to another great good that conflicts with other goods: regularity in the behavior of physical things (often referred to as scientific

102 Plantinga, *God, Freedom, and Evil*, 27–30.

103 John Leslie, "The Theory That the World Exists Because It Should," *American Philosophical Quarterly* 7, no. 4 (1970): 286; John Leslie, "Our Place in the Cosmos," *Philosophy* 75, no. 291 (2000): 20.

104 While some would claim that human actions are partly responsible for certain natural disasters (as through effecting climate change) and for exacerbating the destruction they cause (in the way that cities are constructed, for example), it hardly seems plausible to ascribe all of the harm to free choices.

"natural laws").[105] The extraordinary elegance of the regularities exhibited by objects from quarks to galaxy clusters is not only marvelous in itself but is also indispensable to other great goods, such as the abundant manifestations of beauty in the world. Further, as we explored earlier in discussing the fine-tuning problem, the universe's ordered regularity makes possible the emergence of life, including intelligent creatures like ourselves. It also allows us to make plans and to experience the world as largely comprehensible and predictable.

However, the ordered regularity of the physical world also brings with it things that seem anything but good, like tsunamis, tornadoes, and landslides. Like free will, the good things that the natural laws bring cannot be separated from their undesirable downstream outcomes. As Leslie observes, "A world's causal orderliness could be essential to its having more value than an opium dream, yet causal orderliness is not at all obviously compatible with the absence of all such evils as destruction caused by earthquakes."[106] Thus, the observation of things that seem bad when taken in isolation does not in itself undermine claims that the world's existence is based in goodness.[107]

A related strategy for defending axiological theories from the problem of evil holds that people frequently misappraise the world's goodness due to the limited perspective from which they observe it. None of us can see the cosmos in its entirety. Rather, we have access to only a tiny portion. To correct for this error, we must adopt a more holistic viewpoint that recognizes the value of diversity. Even if the narrow slice of reality within our field of perception at a given time does not seem optimal, it might make a vital contribution to the larger whole that comprises a truly great good. The canvas may possess extraordinary qualities not evident when glimpsing only a miniscule detail. If the components of the universe are more interconnected than is readily apparent, it may be that, as Leslie says, "situations which would be strictly negative in value, each of them worse than a blank, if only they could exist each on its own, could yet make worthwhile contributions to the overall pattern of reality."[108]

Another response to the objection from evil contends that undesirable things are necessary for certain profoundly significant goods, particularly in the form of admirable human actions. In contrast with the inconsistent goods argument, the idea here is not simply that some goods unfortunately bring evils with them, but, rather, that certain evils are integral to the

105 Mawson, *Belief in God*, 205.
106 Leslie, "Our Place in the Cosmos," 20; see also Leslie, *Immortality Defended*, 12.
107 Leslie, *Value and Existence*, 9–10.
108 Leslie, "Our Place in the Cosmos," 21.

flowering of certain goods. Examples include acts of courage and resilience, and the selfless decision to sacrifice personal interests for others' sake. It is not that fear is an unfortunate side effect that follows from acts of heroic courage. Rather, fear is built into the meaning of courage. Exercising courage *is* to act in the presence of fear, and, thus, depends on it. For fear to manifest, there must be something bad to fear. Similarly, a showing of inspiring resilience is by its nature a response to adversity, on which it depends. More generally, in a world devoid of pain or suffering there would arguably be no possibility of acting morally at all, because no actions could harm or benefit others. Morally good acts of unselfishness, honesty, and commitment to doing the right thing would evaporate. Thus, the greater goods contemplated by this kind of argument are conceptually inseparable from the evil that is integrally related to them.[109]

Notwithstanding various attempts to parry the objection from evil, defenders of axiological theories face a difficult challenge due to the sheer magnitude and intensity of the bad things that we perceive in the world.[110] What we see around us does not suggest a vast picture of overwhelming contentment with only the odd blemish here and there. Suffering is far from uncommon, and it strikes capriciously. In addition to the large-scale torment caused by wars and disasters, there are the countless lives that never have a real chance of fulfillment owing to severe poverty or debilitating diseases. Consider as well the number of creatures in the Earth's ecosystem whose survival depends on the consumption of other animals that will perish in agonizing deaths as a result. To some observers, even if arguments defending a teleological cosmos could explain some evils, the staggering quantity of evils cannot be reconciled with a universe that is supposedly directed toward the good.

Living in the Good Universe

In this section we explore an axiological approach's implications for the nature of the cosmos and our place within it. It will be helpful in highlighting the distinctive implications of an axiological approach to begin by drawing contrasts with scientific and mathematical approaches (discussed in the previous two chapters, respectively). As we have noted, in some respects the implications of scientific and mathematical approaches differ radically from one another, due in large part to their opposing views regarding change and contingency. At its core, science studies the nature and causes of change

109 Hick, *Evil and the God of Love*, 324–25.
110 Hume, *Dialogues concerning Natural Religion*, 254–57.

while recognizing—according to the dominant view in physics today—that some events are subject to chance. By contrast, a mathematical approach studies relations that are fixed and necessary, denying the reality of change or chance.

Nevertheless, despite the profound differences, scientific and mathematical approaches also share important features that differentiate them from axiological theories. Scientific and mathematical approaches both identify something at the base of existence that lacks any particular resonance with distinctively human attributes and experience. If the most basic truths concern how physical things behave, then human beings are just another manifestation of those patterns. If the most basic truths concern abstract quantitative relationships, then human beings are just another kind of complex structure with their own necessary mathematical properties. In either case, there is nothing special about human beings from a cosmic vantage point.

On the other hand, an axiological approach identifies something at the base of existence—value or goodness—that resonates with distinctively human preoccupations. The things that matter most to us are not remote in character from the reason for the world's existence. Consider how pervasive the idea of purposes is in our lives. We could hardly recognize our own existence without the idea of directing our actions toward ends. The very idea of purpose is value-laden. It is the value of an end that makes it intelligible as something that guides actions. We pursue ends and, in an important sense, so does the universe. Some people believe that purposiveness is just a product of human invention, but an axiological approach reverses that relation. We exist thanks to ends that are built into the universe's fabric.

Value is central to other quintessentially human concerns. Our experience of the world is filled with sorrows and joys, setbacks and triumphs, nightmare and dreams. We yearn to make sense of our own being. The drive for fulfillment encompasses a desire to understand ourselves in the context of a larger story that hangs together and realizes chosen aims. Axiological theories hold that the cosmos itself is directed toward the good. Value-laden things like fulfillment and meaningfulness have an independent reality. When we make determinations about how to spend our time and resources—including how we treat others—we can think of ourselves as seeking to act in accordance with objective truths. We may conceive of our own ends as informed by or derivative from purpose at the cosmic scale. Moreover, living well means properly orienting ourselves toward universal standards of goodness.

Unlike scientific or mathematical theories of existence, an axiological approach suggests that some things hold a more significant position in the cosmic hierarchy than others. Of course, even if that is so, it does not necessarily follow that human beings are among the things that have a special place. The point, however, is that axiological theories make possible the notion of a special seat at the cosmic table. If goodness guides affairs, then we might be a critical component of the ends toward which events are directed. We are thus free to conceive of life not merely as an intricate combination of quarks and electrons, or a complicated mathematical object, but as uniquely important. We may not live in an utterly indifferent universe after all. If reality itself is end-directed and we are part of its larger purpose, then we could be important characters in the cosmic narrative.

What do axiological theories mean for our ability to comprehend reality? The implications are partly optimistic. We can learn a great deal not only about science and mathematics but also about the realm of value. Truths about value need not contradict or replace the best theories of physics. Instead, there is a division of labor. In proposing theories about the interactions of physical things, science describes *how* the world is, while axiology explains *why* it is. In an axiological theory, ultimate explanation speaks in the language of value. This does not require a return to the Aristotelian categorization of material objects by their final ends. Rather, axiological thought today, as exemplified in Leslie's theory, accepts modern science's study of nature as the playing out of fixed regularities. Nevertheless, goodness enters the picture at a more basic level: it shapes the patterns that govern the natural world in a way that enables it to engender worthy ends (such as cosmic harmony and the development of intelligent life).

However, an axiological approach also implies substantial limitations on our knowledge. Science and mathematics each relies on well-established methodologies that are widely accepted by practitioners within those disciplines. Scientific experiments and mathematical proofs provide grounds for belief that can be recognized by observers with the proper expertise. Yet there is no agreed upon procedure for investigating the good. As discussed above, the seminal axiological thinker, Plato, viewed the attainment of knowledge about the good both as a pursuit of the highest order and as extremely difficult. Indeed, the significance and the difficulty are closely interrelated. It is easy to perceive the particular objects in the world around us with our bodily senses, but this does not provide us with real knowledge. The perfect and unchanging objects of mathematical understanding enjoy a higher status than perceivable particulars, but still do not offer ultimate understanding. The highest truths encompass the highest ends toward which our lives should be directed, which we cannot learn about simply through the use of our bodily senses or the study of mathematics. Attaining better understanding requires reflection, discourse, and an iterative process of questioning one's assumptions.

Consequently, there is a deep tension in Plato's thought regarding the good. On the one hand, there is no higher form of knowledge than that concerning the good.[111] Our deepest yearnings are properly directed toward an enhanced appreciation of the good, which is within our grasp. On the other hand, knowledge of the good is elusive—difficult even for the wisest souls—and no one can expect to attain complete enlightenment.[112] Even if theorists inspired by the Platonic tradition have not always followed Plato himself in every detail, we can see a similar tension in contemporary axiological thought: the realm of value has paramount significance and is open to our understanding, but acquiring knowledge about the good that is certain or complete may forever remain beyond our reach.

Axiological accounts entail another limitation on our understanding, one that cuts to the core of debates over the good as an explanation for the world's existence. The case for an axiological approach rests on a value-laden characterization about the world: that it evinces extraordinary goodness. Our observation of wonderful things in the world is seen as supporting the claim that the very reason for the world's existence is based in the splendid ends toward which it is directed. Yet even if one allows that certain things in the world—like intelligent life—are magnificent, the world also contains many things that do not seem good at all. One need only mention examples such as the last century's two world wars and the innumerable deaths caused by pandemics and other medical afflictions to evoke the kinds of horrors that pose a serious challenge for axiological accounts. A critical challenge for any axiological theory is to reconcile the evidently terrible aspects of reality with an account holding that existence itself is based in goodness.

---

111 It should be noted that Plato's conceptualization of the Good as the ultimate form has been subject to considerable debate among scholars. In Plato's work, most notably in the *Republic*, the Good as the highest form illuminates and gives reality to all other forms. However, interpretations of this claim vary. First, some scholars understand the nature and role of the Good in relation to other forms differently. For instance, some take the Good to possess metaphysical and ontological status for Plato, while others see it as more of an ethical or epistemic principle. Moreover, some philosophers conceive the Good hierarchically, as a kind of "super-form" that grants the other forms their reality, while others suggest that the relationship of the Good to other forms is interdependent, relational in essence, or existing on a continuum. Lastly, the nature of the relationship between the Good and the realm of particulars—and the manner in which the Good imparts "reality" to them—remains much discussed. Some thinkers argue that the Good directly bestows existence upon the physical world, while others take it to function as a guiding principle that physical entities strive toward. Julia Annas, *Platonic Ethics, Old and New* (Cornell University Press, 2018), 96–97, 103–10; Christopher Rowe, "The Form of the Good and the Good in Plato's Republic," in *Pursuing the Good: Ethics and Metaphysics in Plato's Republic*, ed. Douglas Cairns, Fritz-Gregor Herrmann, and Terry Penner (Edinburgh University Press, 2007), 125–26, 130–31, 133.

112 Divine wisdom for Plato is forever beyond human reach (Plato, *Symposium*, 204a–b), since we can at best aspire to be philosophers (or lovers of wisdom) who do not fully attain it. But the redemption of our nature lies in the fact that we can possess a small measure of divine virtue and wisdom. Plato, *Laws*, trans. Tom Griffith (Cambridge University Press, 2016), 906b.

Given the apparently sizable gap between the actual world and a world directed toward goodness, the reconciliation effort must appeal to things that we cannot yet see or fathom. Defenses of axiological theories against the objection from evil stress that we are only able to observe a tiny sliver of reality. Things that appear bad from our limited perspective may nevertheless be a part of a beautiful whole. Another way to shift our perspective is not spatial but temporal: perhaps future events will place apparent present evils in a different light. An axiological approach promises a great deal, including the possibility to gain genuine insights into the reason for the world's existence. At the same time, fulfillment of that promise requires acknowledging current limitations on our ability to fully appreciate how the world that we observe around us today could really be the product of a cosmic drive toward the good.

Questions for reflection and discussion

1. Is human experience inseparable from the idea of value?
2. Is the universe's apparent "fine-tuning" a serious puzzle that needs explaining?
3. Does the idea of a multiverse provide a satisfying answer to the fine-tuning problem?
4. How persuasive do you find Plato's distinction between forms and particulars?
5. Why do you think Plato's ideas about the good have been so influential?
6. What are the strongest arguments that Leslie makes for an axiological theory of existence?
7. Can there be possibilities for existence in the absence of any actually existing things?
8. What arguments best support the objectivity of morality (or value more generally)?
9. Could objective truths about value be independent of the natural world?
10. What is the most significant objection to axiological theories of existence?
11. Which bad things in the world are hardest for an axiological theory to explain?
12. Which responses to the problem of evil are the most promising?
13. What implications would an axiological theory have for humanity's place in the cosmos?

6.

The Divine Universe

Personal Intentions as Explanation

We have thus far considered three types of explanations that are highly familiar to us in our daily lives: scientific causation, mathematical relationships, and the value that different things may hold for us. Every day we navigate the world with a keen awareness of the gravity holding us to the floor, the basic mathematics underpinning the way we make quantitative calculations, and the cherished values motivating our decisions. At the same time, our experience encompasses another kind of explanation that is so pervasive and immediate to the way that we interact with the world that it is easy to overlook. In particular, an event or state of affairs may result from a person's intentional actions.[1] I am directly aware of my self forming intentions and acting upon them. My hand reaches for the door and opens it because I intended that it do so. In this chapter, we will consider an approach to explaining the world's existence that is based in the actions of a personal being who intentionally created the physical world.

The term "God" can take on a wide variety of meanings for different people. For some, God is the divine being described by their particular religious tradition or belief system. For others, God represents whatever they consider to have the greatest significance in the universe or in their lives. God may also stand for the grandest mysteries about the nature of the world

1 Bruce R. Reichenbach, "Explanation and the Cosmological Argument," in *Contemporary Debates in Philosophy of Religion*, ed. Michael L. Peterson and Raymond J. VanArragon (Blackwell, 2004), 100–01, 113.

that stubbornly resist our comprehension. Adopting a view often referred to as "pantheism," some people conceive of God as encompassing virtually everything in the universe itself, while others profess belief in multiple gods. People also differ in regard to the role that beliefs about the divine play in their lives. Divine beings may be the focus of worship, the touchstone for how to live, or a center of communal and cultural life.

Given manifold meanings that people attach to the idea of theism, it is worth emphasizing that this chapter will use the term in a particular way. We are interested specifically in the claim that the reason for the world's existence is to be found in personal explanation: that is, in the intentional actions of a divine being. To be sure, not everyone accepts that personal explanation provides a genuinely distinct kind of reason for events. Some believe that human actions are entirely the product of material processes beyond our control, and that scientific explanations alone can fully account for the way we behave. As we are using the term, a theistic explanation for the world's existence necessarily maintains that intentions can have a reality that is not dependent on the existence of physical things. More specifically, the accounts of theism that we will be exploring insist that at least for one (divine) being, personal explanation has a reality beyond the playing out of scientific natural laws. As an extremely powerful, immaterial being, God can bring things about in a way that does not depend on anything else, and God's intentional action is responsible for the reality of physical existence itself.

Our interest will be specifically in arguments for (and against) theism that are based in reason; that is, in arguments developed through the natural ability of the human mind to draw conclusions from principles considered to be reliable independently of faith or revelation. People have different reasons for their belief that God exists. Some make a commitment of faith to maintain belief in God regardless of the extent to which they find that belief to be supported by evidence or rational argument alone. Belief in God may also be based in words believed to have come directly from the deity. As important as other epistemic grounds are for some believers, we will focus in particular on reasoned arguments bearing on the existence of God.

Our examination will center on three historically influential kinds of arguments for the existence of an all-powerful God: ontological arguments, cosmological arguments, and a style of reasoning known as inference to the best explanation. As we will see, these arguments differ in important ways. Transcending these differences, however, is a core commitment to showing that the reason for the world's existence ultimately rests in the intentions of a divine being.

Ontological Arguments

Ontological arguments are especially intriguing due to the sheer ambition of their claims. They seek to show that God necessarily exists, and they do so entirely with a priori reasoning, meaning that the arguments do not depend on things that we can observe with our senses. Instead of empirical evidence, a priori arguments are rooted in rational reflection. The classic ontological arguments are centered on an analysis of the kind of thing that God is. According to this kind of argument, one can recognize that God must exist just by properly understanding the concept of God. Ontological arguments aim to demonstrate God's existence from indubitable starting points with the deductive rigor that is the bread and butter of mathematics.

One of the most influential ontological arguments was developed by the medieval monk St. Anselm of Canterbury (1033–1109 CE). Anselm drew a distinction between the idea of something that we have in our minds and its actual existence. "[I]t is one thing for an object to exist in the understanding," he wrote, and "quite another to understand that the object exists [in reality]."[2] I might form an idea of a river flowing with caramel and chocolate, but sadly this does not guarantee that it exists in reality. Ordinarily, if someone describes an idea of something that you have never seen, it is appropriate to respond, "Yes, but is there any such thing in reality?" The burden of Anselm's ontological argument is to show that God is an exception in this respect. If you understand the idea of God, you appreciate that God must exist in reality.

Anselm's ontological argument begins with a definition and a crucial premise. The definition of God is "that than which nothing greater can be thought,"[3] and the premise holds that it is greater to exist in reality than to exist in the mind alone. Anselm makes his case in the form of a *reductio ad absurdum* argument (or *reductio*, for short), which assumes the opposite of the conclusion one wishes to support and shows that it generates absurd results. If assuming that X is true leads to something obviously wrong—such as a logical contradiction—then X must be false.

Why would assuming that God does not exist lead to an absurd result? Anselm reasons that if God does not exist in reality, then we can conceive of a being greater than God: namely, a being like God who exists in reality. However, by definition, God is the greatest conceivable being. Thus, to deny God's existence is tantamount to saying that there is a being greater than the greatest conceivable being, which is a contradiction. Therefore, the

2 Anselm, *Monologion and Proslogion with the Replies of Gaunilo and Anselm*, trans. Thomas Williams (Hackett, 1996), *Proslogion* ch. 2.

3 Anselm, *Proslogion*, ch. 2.

assumption that God exists in the mind alone must be wrong. In Anselm's words,

> if that than which a greater cannot be thought exists only in the understanding, then that than which a greater *cannot* be thought is that than which a greater *can* be thought. But that is clearly impossible. Therefore, there is no doubt that something than which a greater cannot be thought exists both in the understanding and in reality.[4]

Gaunilo of Marmoutiers announced a famous objection almost as soon as the ink on Anselm's ontological argument was dry. As a fellow monk, Gaunilo did not question God's existence; rather, the debate concerns the merits of Anselm's argument. To discredit Anselm's ontological argument, Gaunilo's argument employs a *reductio* of its own. In particular, the argument aims to show that if one assumes the soundness of Anselm's argument it leads to an absurd result: namely, the conclusion that there exists in reality any number of other incomparable entities, such as an island greater than which no island can be thought.

To make this "Lost Island" argument, Gaunilo asks us to imagine an island "more excellent than all others on earth," which is "more plentifully endowed than even the Isles of the Blessed with an indescribable abundance of all sorts of riches and delights."[5] Applying Anselm's reasoning, once we have the idea of this most wondrous island in our mind, we must recognize that such an island actually exists. After all, if there were no such island, then we could imagine an even greater island: one that exists in reality. But this is a contradiction, because we began with the idea of the greatest imaginable island. Nor is there any reason to stop at islands; one can with the same reasoning demonstrate the existence of an unsurpassable paper weight, bowl of soup, or mud pile. Since this result is preposterous, the assumption that Anselm's argument is sound must be mistaken.

Although Anselm did not fully explicate a defense of his proof from the Lost Island argument, his reply to Gaunilo suggests the kernel of a rebuttal: that God is unique.[6] God's uniqueness derives from the idea, accepted by many medievals, that there is a hierarchical order of being. Belief in such a hierarchy dates to ancient Greek thought. As discussed earlier, Plato posits two distinct realms of existence—forms and particulars—with the former clearly superior to the latter.[7] Though departing from Plato in many

---

4 Anselm, *Proslogion*, ch. 2.

5 Gaunilo, "Reply on Behalf of the Fool," in Anselm, *Monologion and Proslogion with the Replies of Gaunilo and Anselm*, trans. Thomas Williams (Hackett, 1996), 124.

6 F.C. Copleston, *A History of Medieval Philosophy* (Harper & Row, 1972), 75–76.

7 See Chapter Five.

respects, Aristotle, too, views the cosmos as deeply hierarchical. Aristotle describes rational human beings as positioned higher in the cosmic hierarchy than the beasts but lower than the perfect and incorruptible celestial spheres, which are themselves driven by perpetual yearning toward the being that is the cause of all motion in the universe. Building from Plato's ideas centuries later, the influential Neoplatonist school conceived the One as the apex of reality, whose emanations generated the cosmos's successively lower levels of existence. While the Abrahamic tradition could not fully accept any pagan accounts, it embraced the idea that being was ordered from a supreme being down to angels, humans, and lesser creatures.[8]

The stratified reality that Anselm took for granted fits hand in glove with recognition of a singular, supreme being at the pinnacle of the cosmic hierarchy. Anselm defines God in superlative terms. God is that single being than which none greater can be conceived. There are degrees of existence, and it stands to reason that one being occupies a position above all the rest. This set of assumptions allows Anselm to treat his ontological argument as applying exclusively to God. At the apex of being, God encompasses every possible good characteristic to a maximal degree. Since there can be only one highest being, the same reasoning that leads to the conclusion of God's necessary existence does not apply to Gaunilo's island or to anything else.

Anselm uses his idea of a highest being to further illuminate God's nature. One need only ask which attributes would characterize the greatest possible being. For instance, God must be omnipotent, because if God had less than complete power then it would be possible to conceive a greater being: one with a larger degree of power. Anselm uses similar reasoning to conclude that God is omniscient, omnibenevolent, and eternal. There is a coherence inherent in the very idea of God that is lacking in other ideas, like that of Gaunilo's lost island.[9]

René Descartes (1596–1650) advanced another prominent version of the ontological argument half a millennium later. Descartes offers a different definition of God, but the structure of his argument is similar in important respects to Anselm's. Descartes defines God as a "supremely perfect being."[10] Given this definition, Descartes contends, denying that such a being exists in reality leads to a contradiction. Since a logical contradiction cannot be true, it must be the case that God exists not only as an idea in our minds, but as a being in reality as well. Like Anselm, Descartes assumes that existing is better than not existing, or, in Descartes's terminology, that existence is a perfection. It follows from the definition of God as a supremely perfect being

8 MacDonald, *Being and Goodness*, 102.

9 William E. Mann, "The Perfect Island," *Mind* 85, no. 339 (1976): 417–21, 419.

10 René Descartes, *Meditations on First Philosophy*, trans. Michael Moriarty (Oxford University Press, 2008), Fifth Meditation, 65.

that God possesses every perfection (including omnipotence, omniscience, and omnibenevolence). Since existence is a perfection, God possesses that perfection, too, which is to say that God exists. Existence, then, is not merely accidental to the divine being, but is essential to it, just as having three angles that add up to 180 degrees is an essential feature of what it means to be a triangle. A great mathematician himself, Descartes claims to demonstrate God's existence with the same necessity as a geometrical proof.[11]

Cosmological Arguments

Ontological arguments are not the only way that theists have sought to establish God's existence. Another strategy encompasses a variety of demonstrations collectively known as "cosmological arguments." In contrast with ontological arguments, cosmological arguments are not a priori. That is, unlike ontological arguments, they include among their premises at least one empirical observation about the world. As the term "cosmological" suggests, the observation that serves as the argument's jumping off point is characteristically of a highly general nature, identifying a feature of the universe as a whole. Examples include the observations that events have causes and that things change.

Because ontological and cosmological arguments work differently, some thinkers embrace arguments of one kind while rejecting the other. One of the most prominent proponents of cosmological arguments, Thomas Aquinas (1225–74 CE), rejected a priori demonstrations of God's existence. A Christian theologian who embraced much of Aristotle's thought—including the view that knowledge begins from empirical observations—Aquinas offered multiple versions of the cosmological argument, known collectively as the "Five Ways." We will consider the First Way, an influential version of the argument that Aquinas himself viewed as especially significant.[12] The First Way begins from the empirical observation that some things are in motion. (In Aquinas's Aristotelian parlance, this is similar to saying that things undergo changes.) We will see how Aquinas reasons from this observation to the conclusion that there is an "unmoved mover" (God) that puts other things in motion yet is not itself in motion.[13]

---

11 Descartes, *Meditations on First Philosophy*, Fifth Meditation, 66. As discussed further below, a number of contemporary scholars have developed ontological arguments designed to shore up earlier versions from potential objections by focusing on modal logic (which involves the ideas of contingent, possible, and necessary existence).

12 Rudi Te Velde, *Aquinas on God: The 'Divine Science' of the* Summa Theologiae (Ashgate, 2006), 48.

13 See Thomas Aquinas, "The Treatise on the Divine Nature," trans. Brian J. Shanley, in *Basic Works*, ed. Jeffrey Hause and Robert Pasnau (Hackett, 2014), part I, question 2, article 3.

Working from his Aristotelian roots, Aquinas viewed motion in terms of a relation between potentiality and actuality. Motion is a reduction of something from potentiality to actuality with respect to a particular characteristic. Though Aquinas's discussion is technical at times, we can understand the core idea with the aid of Aquinas's own example. Suppose a piece of wood at a given time is cold. Even though the wood at this moment is not *actually* hot, it has the *potential* to become hot at another time. Now, the movement from a state of potentiality to a state of actuality with respect to a particular characteristic can only be effected by something that is in a state of actuality with respect to that characteristic. Thus, the cold (potentially hot) wood can only become (actually) hot if acted on by something else that is actually hot. Crucially, nothing can be both potential and actual with respect to the same characteristic at the same time. The cold wood is potentially hot only because it is not at present actually hot. Given that something in a state of actuality effects the change while something in a state of potentiality undergoes the change, nothing can both cause and undergo the change at the same time. In other words, objects cannot move themselves.[14] (The cold wood cannot make itself hot.) This means that anything in motion must have been put in motion by something else.[15]

The next step in Aquinas's argument concerns the impossibility of an infinite series of movers. If an object O is in motion, we may ask what moved it. We may next ask about the thing that moved the thing that moved O, and so on. This sequence of questions suggests the possibility of an infinite regress: a series of things where each thing was put in motion by something else in motion without ever reaching an end. Aquinas denies that such an infinite series is possible, because each member in the series of moving things depends on a prior member in the series for its motion. The members in the series only have secondary or instrumental motion in the sense that their motion is derived from another member in the series.[16] But what if we suppose that *every* member in the series *only* had this kind of secondary power? In that case, Aquinas argues, there could be no motion at all. The power of motion cannot be derivative all the way down, because then there would have been nothing to get the series going in the first

14 This principle may seem obvious with respect to inanimate objects, like boulders, but Aquinas says that it also applies to living beings. Plants cannot grow without the Sun, for example, and animals, too, are moved by things around them, as when they respond to the actions of other animals or objects in their environment. Sharon M. Kaye, *Medieval Philosophy* (Oneworld Publications, 2008), 96–97.

15 Edward Feser, *Aquinas: A Beginner's Introduction* (Oneworld Publications, 2009), 65–69.

16 Aquinas makes clear that the sense in which one member in the series is prior to another is not temporal. Although as a devout monk, Aquinas accepted the Christian teaching that God created the world, the First Way is not itself concerned with establishing a first moment in time when motion began. The sense in which God is the first mover is, rather, one of fundamentality. Without God as the beginning of the series that sustains everything else, all motion would cease. *Aquinas*, 69.

place. There must be a starting point of the series that is able to put other things in motion without itself being in motion. Since we know that there is motion—the observation that is the First Way's jumping off point—we can conclude that there is an unmoved mover, and that, Aquinas says, is what we call God.[17] God moves other things, but cannot be moved by anything else.[18]

While the First Way itself does not purport to establish all of God's qualities—such as perfection, immutability, and simplicity, which Aquinas discusses elsewhere—it is meant to establish critical elements of his overall theology. The argument is supposed to demonstrate the existence of God as a being who is distinctive in vitally important ways. Unlike the things for whose motion God is responsible, God is not dependent on anything else for the ability to effect change. God does not represent any kind of potentiality that is reduced to actuality by something outside itself. Rather, God, as the first mover, is pure actuality.[19]

Another of the historically most influential cosmological arguments is of particular interest here, because it is explicitly framed around the question that is this book's central focus. As noted earlier, the eighteenth-century philosopher Gottfried Wilhelm Leibniz employed a distinctively modern formulation when he posed the question: Why is there something rather than nothing?[20]

Where Aquinas's First Way begins from the observation of things in motion, Leibniz's argument begins from the observation that the world contains contingent things. A contingent thing is one that might or might not exist. For contingent things, we might say, existence is optional. We can readily see that things around us are like that. They are subject to change, transformation, and decay. Objects in the world are one way now but they could be different later. They exist at one time but not another. By contrast, a necessary thing is one that must exist; it cannot come into or go out of existence. For a necessary being, non-existence is not on the menu.

Leibniz's argument also employs a crucial premise—the "principle of sufficient reason" (or "PSR")—which holds that everything has a reason for being the way that it is.[21] As Leibniz wrote, "nothing happens without its being possible for him who should sufficiently understand things, to give a reason sufficient to determine why it is so and not otherwise."[22] The principle has been expressed by Leibniz himself and others in a variety of ways, including that all events have causes, that all states of affairs have explanations, and

17 Feser, *Aquinas*, 72–73.
18 Kaye, *Medieval Philosophy*, 98.
19 Feser, *Aquinas*, 73.
20 See Chapter Two.
21 Leibniz, "Principles of Nature and Grace," § 7.
22 Leibniz, "Principles of Nature and Grace," § 7.

that there is an explanation for the existence of every contingent thing.[23] Together, the observation of contingent things and the PSR raise the puzzle of why the world exists. Since the world consists of contingent things, the PSR tells us that there must be something that explains its existence.

Our immersion in contingent things makes the fact of their existence easy to take for granted. What could be more familiar than our own existence and the existence of everything around us! Nevertheless, Leibniz wanted us to feel puzzled by it. The contingent things that surround us did not have to exist. It seems, then, that there could have simply been nothing at all. Yet the world of contingent things exists, and that fact cries out for explanation. We should feel awe and wonder that absolute nothingness is not the prevailing state of affairs. To further provoke the appropriate astonishment, Leibniz claimed that absolute nothingness would have been the simplest and easiest possible state of affairs. Why, then, is there something rather than nothing? While the question may seem daunting or elusive, the PSR assures us that the question has an answer; there must be an explanation for the world's existence.[24]

Leibniz's argument next reasons about the nature of an answer to this profound question. In seeking to explain one particular contingent thing, we find ourselves referring to something else—another contingent thing—but that contingent thing, too, requires an explanation. A table was built from the wood which came from a tree, which grew from a seed, and so on. However, since a contingent thing might or might not exist, its existence cannot be accounted for simply by pointing to another object that might or might not have existed. If the explanation for each contingent thing is another contingent thing, Leibniz thinks, we ultimately have no explanation at all. It follows that the explanation for contingent things cannot lie in a series of other contingent things, even if that series includes the entire collection.

Leibniz's argument uses the insufficiency of contingent things as a satisfying explanation to establish the existence of a necessary being. The PSR tells us that there is a reason for the existence of contingent things, but contingent things cannot explain themselves. The world of contingent things, then, must find an explanation in something outside itself: something which does not depend on something outside itself for an explanation of its existence. Thus, the stopping point for this series of explanations must be a necessary being, one that can explain its own existence, and, thus, the existence of all contingent beings. A necessary being is one for whom existence is an integral aspect

23 Daniel O. Dahlstrom, "Being and Being Grounded," in *The Ultimate Why Question: Why Is There Nothing at All Rather Than Nothing Whatsoever?*, ed. John F. Wippel (Catholic University of America Press, 2012), 128.

24 Leibniz, "Principles of Nature and Grace," §§ 7–8.

of its essence. As Leibniz put the point, we do not have to look outside the necessary being for an explanation of its existence, because it carries "the reason of its existence within itself."[25] God is the necessary being that answers the question why there is something rather than nothing.[26]

Notwithstanding differences in the empirical observations and premises on which they rest, Leibniz's argument shares a similar overarching structure with Aquinas's First Way (and many other cosmological arguments). Together with the observation of contingent things, the PSR initiates an explanatory regress in the sense that everything in the world must find an explanation in something else. However, the individual things in the world—whether taken one at a time or as a collectivity—cannot provide a satisfactory explanation, because they simply point back to things of the same general kind: things which are not capable of providing a solid foundational point. The only way to explain the world of contingent things is by appealing to something that is outside of and fundamentally distinct from it. Only recognizing the existence of God—a being not dependent on anything outside itself to explain its existence—can explain the observation with which the argument begins: the existence of contingent things. In broad structure, the argument shares similarities with Aquinas's First Way, which tells us that only something that is outside of and fundamentally distinct from the things around us can explain the observation with which that cosmological argument begins: that objects are in motion.[27]

God as the Best Available Explanation

Another kind of argument for the existence of God is known as "inference to the best explanation" (or "abductive argument"). Broadly speaking, abductive arguments—which play important roles in a wide variety of contexts—comparatively assess proposed explanations according to a set of pre-established criteria. The relevant factors typically include things like

---

25 Leibniz, "Principles of Nature and Grace," § 8.

26 Like Anselm and Descartes, Leibniz thought that he could not only demonstrate God's existence but also infer God's attributes. For instance, God must be immaterial, because only a being with a fundamentally distinct nature from the world could be responsible for it. Moreover, to have created the entire world, God must be a being of the greatest power. Since the necessary being specified this world from all of the possibilities, we can further conclude that God possesses unlimited understanding and the ability to will things into existence. William Lane Craig, *The Cosmological Argument from Plato to Leibniz* (Macmillan, 1980), 275.

27 Graham Oppy, *Arguing about Gods* (Cambridge University Press, 2006), 100. Oppy discusses similarities among Aquinas's First Way (involving the origin of motion, which we have discussed) and Third Way (which, though not identical to Leibniz's argument, is also based in the distinction between necessity and possibility). See also P.T. Geach, "Aquinas," in *Three Philosophers*, ed. G.E.M. Anscombe and P.T. Geach (Basil Blackwell, 1963), 111–14 (discussing Aquinas's First Way).

how well an explanation accounts for the phenomena under consideration, and an explanation's consistency with other firmly held beliefs. With respect to the question under consideration, an abductive argument aims to show that a particular theory does a better job of explaining things than anything else. An abductive argument for the existence of God, then, is supposed to establish that recognizing God's existence provides a more effective answer to certain questions than any other available answers.

Although those employing abductive analysis do not always use the term or spell out the steps explicitly, this style of reasoning is pervasive, from everyday reasoning about the likely causes of observable circumstances to the practice of science. For example, observing puddles on the ground in the morning might lead me to conclude that it rained the night before. Abductive reasoning is also common in medical care. Doctors do not typically diagnose patients through deductive reasoning, drawing logically necessary conclusions from initial premises. Instead, they seek the best available explanation for a set of observations.

Consider a doctor facing the question of why a patient has suddenly acquired three different symptoms. One possibility is that the patient contracted a single disease (D1) that is known to cause all three of the symptoms. However, the doctor is also aware of three other diseases (D2, D3, and D4), each capable of causing one of the three symptoms. All other things being equal, which is a better explanation: that the patient has contracted disease D1, or that the patient has contracted all three of the other diseases? The doctor is likely to reason as follows: "One criterion that helps to assess competing explanations for the same set of observations is their relative simplicity. Since it is simpler to suppose that the patient contracted only one disease, that explanation is preferable."

Abductive reasoning is widespread in science more generally. Take the question of why dinosaurs became extinct. Many experts believe that debris in the atmosphere occasioned by Earth's collision with an asteroid altered the climate to such an extent that most species could not survive. The basis for this belief is not step-by-step necessary inferences in a deductive proof. Other explanations are logically possible. Instead, the persuasiveness of the asteroid hypothesis lies in the comparative application of criteria to proposed explanations. This means that we might revise our beliefs in light of new evidence. Suppose fresh empirical observations suggest that it is extremely unlikely that a large asteroid could have struck the Earth during the dinosaurs' reign. This information would weaken our confidence that the asteroid hypothesis was still the best explanation.

Some recent thinkers have developed detailed, systematic, abductive arguments in support of the claim that God is the best explanation for the world's existence. The work of contemporary philosopher Richard Swinburne (b. 1934) is especially interesting due to its emphasis on the use of criteria for evaluating explanations that are widely accepted by scientists.[28] In general, abductive analysis involves several key elements: (1) a question, or something that demands explanation; (2) criteria for assessing proposed explanations; (3) candidate explanations; and (4) comparative assessment of the explanations according to established criteria. We will consider each component of Swinburne's argument in turn.

The question that initiates Swinburne's analysis is the same as ours: Why does the world exist? Like us, he is seeking an ultimate explanation, which he describes as "that object or objects on which everything else depends for its existence and properties." Whatever this object or objects turn out to be, it will account for "everything observable."[29] We might think of this criterion as the job description for such an object, and abductive analysis as a means of assessing the applicants.

At the outset, Swinburne places a constraint on the nature of ultimate explanation. In particular, he insists that the ultimate explanation must be a brute fact: something that just is and has no explanation. The ultimate explanation—whatever it is—will explain everything except itself.[30]

Why must the ultimate explanation be a brute fact?

Swinburne argues that if the ultimate explanation were not a brute fact, then it would have to be either self-explanatory or logically necessary, and he rejects both possibilities. He rules out self-explanation as circular, because only something that already exists is capable of serving as an explanation. Thus, to claim that something is self-explanatory means assuming its existence from the start, which would not provide a satisfying answer to the question of why there is a world. Swinburne's argument that the ultimate explanation cannot be logically necessary is based on two assumptions. First, anything explained by something that is logically necessary must also be logically necessary. Since the ultimate explanation by definition explains everything, a logically necessary ultimate explanation would imply that everything in the universe is logically necessary. However, this would contradict Swinburne's second assumption: that some things in the world—such as free choices made by people or God—are not logically necessary.

Swinburne's case for God is not only abductive but also teleological. As discussed earlier in the context of axiological theories, teleological arguments support their conclusions by claiming that they best explain the

28 Richard Swinburne, *The Existence of God* (Clarendon Press, 2004), 136.

29 Richard Swinburne, *Is There a God?* (Oxford University Press, 1996), 39.

30 Swinburne, *The Existence of God*, 73.

existence of certain good things in the world.[31] Teleological arguments commonly emphasize the existence of living organisms—and especially intelligent creatures like human beings—as a remarkably good thing. While Swinburne is no exception in this regard, he also emphasizes a good feature of the world that he considers underappreciated: the orderliness of nature on which life depends. Consider the widely accepted scientific belief that every electron in the universe behaves according to precisely the same patterns. Each of the other elementary particles—neutrinos, muons, the various kinds of quarks, and so on—is also assumed to have its own patterns of behavior that apply to every one of its kind. That is a mind-boggling number of entities all dancing to the same drummer. Moreover, these regularities do not apply only to different particles of the same type at a given moment in time or local region in space. They are believed to obtain for all time and across the entire universe. Since our own existence depends on these unvarying patterns of behavior, Swinburne views the regularity of nature as a wonderfully good feature of the world that cries out for explanation.[32] The explanation of the world's existence, then, must also explain why the world contains remarkably good features.

Having presented the central question, Swinburne provides the criteria that he thinks generally apply in assessing explanations: (1) simplicity, (2) accuracy (whether it makes correct predictions), (3) surprise (whether it accounts for things that we would not otherwise expect), and (4) conservatism (whether it fits with other accepted beliefs about the way the world is).[33] However, Swinburne thinks that the last criterion—conservatism—does not apply to explaining the world itself, because in that context nothing counts as pre-established background knowledge. With conservatism eliminated as a criterion, Swinburne frames his inquiry as a quest for the simplest theory that accurately predicts surprising phenomena.[34]

Among these criteria, Swinburne accords the highest value to simplicity—which he calls "the sign of the true"[35]—noting its critical role in science. One reason that simplicity plays an important role in science is that there are always multiple scientific theories that would predict or explain the same set of observable phenomena. Scientists rely on simplicity as a basis for favoring some theories over others.[36] But what makes one theory simpler than another? The meaning of simplicity is not self-evident, and people interpret the concept differently. For Swinburne, the simplest explanation is

31 See Chapter Five.
32 Swinburne, *Is There a God?*, 50–51.
33 Swinburne, *Is There a God?*, 25–26.
34 Swinburne, *Is There a God?*, 25–32.
35 Swinburne, *The Existence of God*, 59.
36 Swinburne, *The Existence of God*, 58–61.

the one that posits (or assumes the existence of) the fewest things (whether properties, objects, or governing principles).[37] Since Swinburne claims that the ultimate explanation cannot itself be explained, this means that at least one unexplained (or brute) fact must be posited. Thus, Swinburne seeks the ultimate explanation that posits the fewest brute facts that successfully explain everything else.

With the central question and the criteria for assessing proposed answers established, Swinburne next describes the candidate answers. Swinburne's three candidates derive from his assumption that there are fundamentally only two types of explanations: scientific and personal. By "scientific explanation," Swinburne means an explanation that answers a question about why something occurred by referring to the properties of physical objects—their patterns of behavior—which he calls their "powers and liabilities."[38] This is similar to what we have in mind when saying that one physical event caused another.

By contrast, a "personal explanation" is rooted in the intentional actions of a rational agent.[39] For Swinburne, personal explanation is genuinely distinct from scientific explanation, meaning that persons can be causes of events in a way that is not captured merely by describing the physical properties of their bodies.[40] Some events have non-physical causes.[41] Like God, human beings can initiate actions in a non-physical manner through the formation of intentions. But there is a difference. While people can *initiate* an action (such as the raising of a hand) by forming the intention to do so, we can only *complete* the action through the physical movements of our bodies. For example, I can initiate the processes leading to the opening of a jar by forming the intention to reach for the jar, but I cannot actually open it without the operation of my physical body with its bones, muscles, nerves, and so on. By contrast, as an immaterial being, God can make things happen directly through personal explanation without the mediating mechanisms of a physical body.[42]

Based on these two types of explanation—scientific and personal—Swinburne identifies three candidate explanations for the world's existence. The first option, *materialism*, finds ultimate reasons exclusively in scientific explanation. The second option, *theism*, finds ultimate reasons entirely in personal explanation: specifically, the intentions of a single divine being. The third option, *humanism*, relies on scientific explanation to explain the

37 Swinburne, *The Existence of God*, 53.
38 Swinburne, *Is There a God?*, 39–40.
39 Swinburne, *The Existence of God*, 43–44.
40 Swinburne, *The Existence of God*, 97.
41 Swinburne, *The Existence of God*, 41–44.
42 Swinburne, *The Existence of God*, 40.

behaviors of physical objects, while depending on personal explanation to explain how human beings initiate actions through the formation of intentions. Since humanism recognizes the independent reality of personal explanation—but not the existence of God—it entails a mix of both scientific and personal explanation.

Swinburne's argument analyzes the relative simplicity of the three candidate explanations. Recall that Swinburne defines simplicity according to how many things an explanation must posit. Theism wins the contest, because it must posit only one thing. Materialism must posit an unfathomably large number of things, and humanism is even worse than materialism on this score.

Swinburne dispatches with humanism quickly on the grounds that it must posit everything materialism does plus each human being's ability to act through the exercise of personal intentions.[43] On the other end of the spectrum, theism must posit only one brute fact: the existence of God. Swinburne's defense of this claim begins with a definition of God as the simplest possible personal being. As a personal being, God possesses all of the properties that persons have to at least some degree. From here, Swinburne moves to the conclusion that God has all of the properties that persons have to an *infinite* degree. He reaches this conclusion on the back of a crucial premise: that the simplest properties are those that a being either possesses to an infinite degree or does not have at all. That is, when it comes to properties, all or nothing is simplest. (Swinburne notes that scientists have long preferred to hypothesize entities that manifest properties either to an infinite degree or not at all.)[44] But if God possesses all the properties of persons to a non-zero degree, and these properties are as simple as can be, then God must possess them infinitely. Thus, Swinburne argues, theism does not even have to separately posit God's various attributes, because they all follow from the positing of God as the simplest personal being.

For example, since persons by their nature possess at least some power, knowledge, and freedom, God must be omnipotent, omniscient, and maximally free. Maximum freedom, in turn, means that the only cause of divine actions is God's own choices, because that is a simpler manner of acting than if God had to act in a way involving outside causes.[45] Similarly, God is eternal, because otherwise additional factors would be needed to explain the times when God did not exist.[46] Thus, given the single brute fact of God's existence, the divine attributes necessarily follow.

43 Swinburne, *The Existence of God*, 106; Swinburne, *Is There a God?*, 41.

44 Swinburne, *The Existence of God*, 96–97; Swinburne, *Is There a God?*, 44.

45 Swinburne, *The Existence of God*, 98.

46 Swinburne, *Is There a God?*, 45–46.

Swinburne's teleological account also includes a case for God's goodness. The argument rests on a number of premises: (1) that as a free person—a rational agent—God acts for reasons; (2) that to act for reasons is to pursue the best possible state of affairs; (3) that there are objective truths about what is good; and (4) that a genuinely free being pursues the good. Together, Swinburne's assumptions mean that God wishes to act for the good, has complete knowledge of the good, and has the power to realize the good. Always acting for the best, God creates the world because it is good. In Swinburne's words: "God chooses for reasons, or between reasons, and brings about the universe because it is one of the many good things he could bring about."[47]

But why should we think that a perfectly good God would create a world like ours? We see good things around us, Swinburne says, that would be difficult to explain without assuming that the world was designed with them in mind. The world is ordered according to patterns that are inherently beautiful. It is populated with a marvelous variety of animals,[48] including one species of intelligent creatures who make choices, ask questions, learn, and even develop technologies to enhance the things around them.[49] Here, Swinburne invokes the fine-tuning problem, observing that it is extremely unlikely that the conditions needed for the development of human life could have come about solely by chance.[50] In sum, Swinburne maintains that the simplest possible being created the world because of its goodness.

Materialism, on the other hand, scores extremely poorly on the criterion of simplicity. A materialist explanation must separately posit the existence and properties of each individual physical entity: every electron, muon, quark, and so on. That is an immense number of brute facts. Moreover, Swinburne asserts, materialism offers no satisfactory response to the fine-tuning problem. It has no explanation for why the world turned out to have features consistent with the emergence of intelligent life against such staggering odds.[51] Darwin's theory of natural selection explains how complex life arises from less advanced forms, but says nothing about why the preconditions for the emergence of life were satisfied in the first place.[52] By contrast, theism shows that an all-powerful God intentionally created a world with features that make creatures like ourselves possible.

Notwithstanding Swinburne's rejection of materialism, he is at pains to show that theism is consistent with science. His aim is not to displace the

47 Swinburne, *Is There a God?*, 47.
48 Swinburne, *Is There a God?*, 62–63.
49 Swinburne, *Is There a God?*, 52–54.
50 See Chapter Five.
51 Swinburne, *Is There a God?*, 23–24; 49–50.
52 Swinburne, *Is There a God?*, 60–61.

role of science, but, rather, to explain its foundation. He observes: "The very success of science in showing us how deeply orderly the natural world is provides strong grounds for believing that there is an even deeper cause of that order."[53] If the Big Bang Theory is true, this involves no contradiction of Swinburne's account,[54] because his theism is not meant as a cork to plug the holes in science. Instead, theism lies beneath as a more fundamental explanation for the things that are true about science. As Swinburne puts it: "I am postulating a God to explain what science explains; I do not deny that science explains, but I postulate God to explain why science explains."[55]

Challenges for a Theistic Approach

We turn now to challenges for a theistic approach to the puzzle of existence. Given the vast scope of debates over God's existence spanning millennia, we will focus selectively on a few especially prominent objections to theism. We begin with objections to particular arguments seeking to demonstrate the existence of God, before considering more general arguments that theism is incoherent or highly implausible.[56]

We have already considered Gaunilo's Lost Island objection to Anselm's ontological argument. Here, we discuss an influential objection advanced by the Enlightenment philosopher Immanuel Kant (1724–1804), which targets Descartes's version of the ontological argument. Recall that Descartes defines God as a supreme being who possesses every perfection. If existence is a perfection, God must possess it, which is to say, God exists. Although Kant was a theist, he denied that mere reflection on a concept could demonstrate God's existence.

In itself, there is nothing remarkable about drawing necessary conclusions from the definition of a concept. That kind of analysis is familiar in mathematics, for example. Given that a triangle is a polygon with three sides, it follows necessarily that anything fitting this definition must have

53 Swinburne, *Is There a God?*, 68.

54 Swinburne, *Is There a God?*, 60–61.

55 Swinburne, *Is There a God?*, 68.

56 Some objections to theism center on issues that have attracted a tremendous amount of philosophical attention entirely apart from debates over theism. Examples include whether free will exists and whether personal explanation represents a distinct kind of explanation from those offered by scientific theories. Since theists commonly claim that God freely chose to create the world, it poses a potential problem for theism if free will or genuinely free choices are impossible. Similarly, it poses a threat to theism if there is no such thing as personal explanation. Theism as an ultimate explanation views physical existence as resulting from the intentional actions of an immaterial person. That kind of view is called into doubt if personal explanation in the end is nothing more than the product of physical processes like those that are the subject of scientific explanations. As interesting and important as these questions are, we will focus principally on debates that are waged more specifically in the context of debates over theism.

angles adding up to 180 degrees. To use a more fanciful example, if we define a "frepper" as a table made of wood that is painted entirely in purple, then it follows with logical necessity that no freppers are made of plastic or have yellow legs. These conclusions are rooted in the relation between different properties associated with an idea. We infer that if an object is defined as possessing property X, then it must also possess some other property Y.

Why, then, did Kant object to the ontological argument? In calling existence a perfection, Descartes treated existence as an attribute—or property—that things have. However, Kant denied that existence was a property (or "real predicate," as he put it). As Kant wrote, "Being is obviously not a real predicate; that is, it is not a concept of something which could be added to the concept of a thing. It is merely the positing of a thing, of certain determinations, as existing in themselves."[57]

A property normally tells us more about what an object is like. When we learn about an object's properties, we adjust our concept of what it means to be that kind of object. Suppose I have an idea of a particular object and you tell me that something fitting that description exists in reality. In doing so, you have not given me any new information about the kind of object that I have in mind. You have not enhanced my idea of what it is. If I have an idea of a frepper as a purple table made of wood and someone tells me that there is a frepper in a store downtown, I do not thereby change my understanding of the term (the way it would, for example, if you informed me that all freppers have four legs). About any object defined in a particular manner, we may inquire whether anything like that actually exists. This existential question is not about the properties that are inherent in the idea. It is an empirical question about the state of affairs that obtains in reality. Descartes's argument treats existence as a property. But if existence does not even count as a property, then it cannot be the subject of a necessary relation between two or more properties (as when Descartes says that a supremely perfect being must have the property of existence).

More recently, a number of philosophers have proposed versions of the ontological argument that claim to avoid Kant's objection. These newer arguments largely focus on modal logic, which concerns the ideas of impossibility, contingency, and necessity. Philosopher Norman Malcolm (1911–90), for example, recognizes that *contingent* existence is not a perfection, but he contends that *necessary* existence *is* a perfection.[58] Contingently existing things depend on other things for their origin and continuance. They may cease to exist and might never have existed. In Malcolm's view, that kind of

57 Immanuel Kant, *Critique of Pure Reason*, trans. Norman Kemp Smith (St. Martin's Press, 1963), A598.

58 Norman Malcolm, "Anselm's Ontological Arguments," *Philosophical Review* 69 (1960): 46.

existence—shared by mundane things like shards of glass, twigs, and soda cans—can hardly be considered a perfection. Necessary existence, however, is another matter. If something exists necessarily, it does not depend on anything else for its existence and cannot cease to exist.[59]

For Malcolm, necessary existence, unlike contingent existence, qualifies as a property and can be treated as a perfection. If we learn that something exists necessarily, we have learned something new about the kind of thing that it is, and we have learned of a property that makes the thing greater than it would be otherwise. According to Malcolm, if God is understood in terms that include the property of necessary existence, the ontological argument is sound. With God defined as possessing all perfections—and necessary existence recognized as a perfection—we can say that if God's existence is possible, then God possesses the quality of necessary existence, which means that God actually exists. Without a reason to think that God's existence would entail a contradiction, Malcolm concludes that God's existence is possible, leading to his conclusion that God must exist in reality.[60]

Philosopher Alvin Plantinga (b. 1932) similarly bases his ontological argument on necessary existence. He draws heavily on the idea of possible worlds, an approach to thinking about modal logic that has received a great deal of attention from philosophers in recent years. Plantinga is not fully satisfied with Malcolm's argument, which he thinks shows only that every possible world contains a maximally great being, but not that this being has all of the qualities commonly associated with God. To shore up the argument, Plantinga introduces the idea of maximal excellence.[61] A maximally great being is one that not only exists necessarily (that is, exists in every possible world) but also possesses the properties of omnipotence, omniscience, and perfect goodness in every possible world. Plantinga further posits that there is a possible world in which a maximally great being exists (which is similar to saying that the existence of a maximally great being is possible). Once we say that a maximally great being—God—exists in a possible world, and that such a being by definition exists in *every possible world*, it follows that God exists in our world.[62]

While many recognize the sophistication of recent versions of the ontological argument, questions remain about their persuasiveness. One objection to Plantinga's argument, for example, challenges the premise that a maximally great being exists in at least one possible world. Without that

59 Malcolm, "Anselm's Ontological Arguments," 46–47.

60 Malcolm, "Anselm's Ontological Arguments," 50–51.

61 Brian Davies, *An Introduction to the Philosophy of Religion*, 4th ed. (Oxford University Press, 2021), 122–23.

62 See Plantinga, *The Nature of Necessity*, 196–221; Plantinga, *God, Freedom and Evil*, 85–112.

premise, the argument never gets off the ground.[63] Plantinga himself has acknowledged the potential vulnerability of his argument to this kind of challenge, conceding that the argument does not "prove or establish" God's existence, but only shows that "it is rational to accept" belief in God.[64] The debate highlights the significance of whether any particular definition of God is coherent and describes a being who could possibly exist, a question that we further explore below.

Next, we consider objections to the cosmological arguments advanced by Aquinas and Leibniz, respectively. One basis for challenging Aquinas's First Way is to reject one or more of its starting points. Aquinas's First Way begins from the uncontroversial observation that some things are in motion. However, it also rests on more contestable premises. Aquinas assumes both that the motion of any object in motion can be explained in terms of a cause that is found outside of itself, and that there cannot be an infinite regress of causes for motion. From this he concludes that God is the first mover: something that is responsible for the motion of other things without itself being put in motion by anything else. Critics have challenged both of Aquinas's assumptions.

First, some reject the claim that motion cannot occur without a cause, or without a cause outside of the thing that is in motion. There seem to be some things that are capable of moving themselves. On certain understandings of free will, for example, human beings can do things of their own accord, as when we change our minds, or initiate an action by exercising our own personal agency.[65] Some also view certain physical movements—such as an atom's emission of radiation—as expressions of their intrinsic qualities (rather than as having been caused by something external). Others contend that motion does not require a cause at all. Contemporary claims of this kind draw on scientific understandings unknown to Aquinas. For example, Newtonian physics holds that bodies in motion stay in motion unless acted upon by an outside force.[66] Moreover, according to prevalent interpretations of quantum theory, events occur at the subatomic level that are genuinely random, with no explanation for why a particular event occurred in one way rather than another.[67] Indeed, some view the origin of the universe—the Big Bang—as itself an uncaused event.[68]

63 Graham Oppy, "Ontological Arguments," in *The Stanford Encyclopedia of Philosophy*, ed. Edward N. Zalta and Uri Nodelman (2023), § 8, accessed December 10, 2024, https://plato.stanford.edu/archives/fall2023/entries/ontological-arguments/.

64 Plantinga, *The Nature of Necessity*, 221.

65 Oppy, *Arguing about Gods*, 103.

66 Anthony Kenny, *Medieval Philosophy* (Oxford University Press, 2005), 303.

67 See Chapter Three.

68 Oppy, *Arguing about Gods*, 103.

A good deal of debate over the First Way has also focused on Aquinas's assumption that there cannot be an infinite regress of causes for motion. Disagreement over the acceptability of infinite regresses in explanations dates to ancient times, and the subject remains controversial. Some people accept Aquinas's position on regresses as highly intuitive. Others support it by arguing that observed motion cannot be the product of an infinite chain of movers because it would be impossible to traverse such an unending series.[69] Aquinas's critics, on the other hand, object that so long as each instance of motion can be explained in terms of a prior motion, we have no basis for demanding anything more.[70]

In response to this objection, Aquinas distinguishes between two kinds of series. One kind of series is ordered *sequentially*. The items in a sequentially ordered series have a linear temporal relation to one another. One thing causes another, which causes another, and so on. In this kind of series, once something is in motion it has the independent ability to cause motion in something else. For example, a person has children, who grow up and have children of their own. The ability of the grandchildren to procreate does not depend on the continued existence (or motion of) the grandparents. With respect to *this* kind of series, Aquinas says, there is in principle no reason why there cannot not be an infinite regress of things in motion.[71]

Another kind of series (at issue in the First Way) is ordered *essentially*. The items in an essentially ordered series are not arrayed in a linear timeline, and they do not have an independent ability to effect motion. Instead, each item's power to effect motion depends on the prior item. Its power is purely instrumental. Aquinas uses the example of a staff moving a stone which is moving a fallen leaf. The items are moving together at the same time, yet all of the motion would cease without the staff's motion. This is the kind of series that Aquinas has in mind when he asks why there is any motion at all. In contrast with a sequentially ordered series, an essentially ordered series must have a first mover. The other items are only members of the series by virtue of their relation to the first mover. Aquinas embraces the Christian teaching that God created the world, but the First Way is not itself concerned with establishing a first moment in time when motion began. Rather, the argument is meant to show that God is primary among movers in terms of fundamentality. Without God as first mover, there could be no motion in the first place.[72]

---

69 Sharon M. Kaye, *Medieval Philosophy* (Oneworld Publications, 2008), 98.

70 Oppy, *Arguing about Gods*, 101–03.

71 Feser, *Aquinas*, 69–70. Although Aquinas believed that there was not in fact an infinite series of sequentially ordered causes of motion, the important point here is that he did not think that this could be established one way or the other purely through philosophical reasoning. Feser, *Aquinas*, 71–73.

72 Feser, *Aquinas*, 69–73; Oppy, *Arguing about Gods*, 101.

Debate over Aquinas's First Way reflects disagreements over what counts as an acceptable explanation. Many of Aquinas's critics say that we get sufficient explanations about the world from science. Motion can be adequately understood through the study of physical interactions. Aquinas, however, denies that science alone provides satisfying explanations for basic features of the world, such as the fact that things are in motion. Explaining motion depends on something fundamentally different from the objects studied by empirical science. Objects cannot move themselves, and nature cannot explain itself. Understanding the whole of being requires a distinctive plane of explanation. In the language of Aquinas's Aristotelian metaphysics, only a being who is pure actuality (God) can explain the series of motions by which potentialities are reduced to actuality. We must transcend the things studied by science to understand how any motion could be possible at all.[73]

Prominent objections to Leibniz's cosmological argument share broad similarities with objections to Aquinas's First Way. Leibniz's argument begins by observing that the world consists of contingent things. Leibniz then assumes that there is an explanation for the existence of contingent things, and holds that a collection of contingent things cannot explain its own existence. He therefore concludes that there is a necessary being (God) who explains everything else while requiring no external reason for its own existence. Earlier, we considered the criticism of Aquinas's First Way that Aquinas has no basis for rejecting an infinite regress in causes of motion. In a similar spirit, Leibniz's critics claim that there is no reason for rejecting a chain of explanations where each contingent thing is explained by another contingent thing. If each contingent thing has an explanation, the argument goes, we have no reason to demand an explanation for the whole collection of contingent things. As one scholar writes, "As long as the presence of each contingent member of the set ... is explained, *ipso facto* the whole set is explained. There's no fact 'left over' ... needing an explanation."[74] Thus, Leibniz has no basis for insisting that only God can explain the existence of contingent things.[75]

Debate about cosmological arguments raises questions about whether explanations for the world's most basic features must transcend science. Aquinas and Leibniz contend that we can only explain certain aspects of the world by appealing to something fundamentally different from the natural world studied by science. Objections to Aquinas's and Leibniz's arguments, on the other hand, maintain that things in the world itself provide the only kinds of explanations that we need or are justified in demanding.

73 Te Velde, *Aquinas on God*, 51–54.

74 Paul Edwards, "The Cosmological Argument," in *Readings in the Philosophy of Religion*, ed. Baruch Brody (Prentice-Hall, 1974), 76–77.

75 Oppy, *Arguing about Gods*, 120; Mawson, *Belief in God*, 156.

A related objection charges that Leibniz illegitimately excuses God from requirements that apply to everything else. (We may call this the "illegitimate exemption" objection.) According to the PSR, everything must have an explanation, but Leibniz then finds a stopping point for explanation in God, who needs no outside reason for existing. The illegitimate exemption objection complains that Leibniz uses the PSR to drive the argument but then arbitrarily abandons it at the last minute. The nineteenth-century philosopher Arthur Schopenhauer wrote that cosmological arguments like Leibniz's treat the PSR's requirements "like a cab which we dismiss after we reach our destination."[76] Leibniz relies on the PSR to establish God's existence, but then excuses God from the PSR's requirements.

In responding to the illegitimate exemption objection, Leibniz draws a distinction between contingent and necessary existence.[77] The PSR guarantees explanations for things that exist *contingently*. Contingent things are things that might have been different than they are and, thus, Leibniz assumes, there must be a reason why they turned out one way rather than another. God, though, is not contingent. God is a necessary being whose existence contains its own explanation; it could not have been other than it is. Thus, it is consistent with the PSR to say that God's existence does not require an explanation from anything outside itself the way that contingent beings do.[78]

Leibniz's argument depends on the idea of a particular being existing necessarily. Earlier, we discussed Kant's denial that the idea of necessity properly applies to the fact of a thing's existence. The coherence of necessary existence is one of many questions that Leibniz's cosmological argument raises regarding the nature of existence and the requirements of satisfying explanations. Perhaps none of these questions is more fundamental than whether we should accept Leibniz's PSR as a sound premise.

Consider the overarching question of how we should decide what to believe. The question reflects back on itself. Deciding what to believe inevitably draws on pre-existing beliefs: in particular, beliefs about what qualify as good reasons for accepting propositions as true. No question about the grounding of our beliefs is more basic than when we should expect things to have reasons in the first place. The PSR presumes that there are reasons for all events, states of affairs, and existing things (or at least for those of a contingent nature). But why should we believe *that*?

---

76 Arthur Schopenhauer, *On the Fourfold Root of Sufficient Reason*, trans. Karl Hillebrand (George Bell and Sons, 1903), 43.

77 Alexander R. Pruss, "A Restricted Principle of Sufficient Reason and the Cosmological Argument," *Religious Studies* 40 (2004): 172.

78 Leibniz, "Principles of Nature and Grace," § 8.

The question's profundity makes it difficult to address with non-circular arguments. Assessing the PSR hinges on prior assumptions about the role of reasons in our thinking. It is hard to see how one could demonstrate the PSR's truth or falsity. It does not seem to be empirically testable, for example, or required (or prohibited) by principles of logic.

Nevertheless, we can recognize appealing aspects of accepting the PSR. It may be tied to our self-conception as rational creatures. The pursuit of explanations is arguably the sine qua non of rationality itself.[79] Moreover, many think of *reasonableness* as an essentially human characteristic. If so, one might wonder: what does reasoning mean if it does not include the pursuit of *reasons* for things being as they are? We might view the PSR as constitutive of the kinds of rational creatures that we take ourselves to be.[80]

Some thinkers, however, deny the PSR's indispensability. Even if seeking reasons is essential to the kinds of creatures that we are, they argue, that does not necessarily mean that there must be a reason for *everything*. Moreover, contemporary science suggests that some events lack explanations. The prevailing view among physicists is that certain physical events occur in a manner that is genuinely random.[81] At the same time, a prior commitment to the PSR furnishes a basis for concluding that the prevailing view of physicists must be wrong. For PSR supporters, recognizing any special exceptions from the principle may seem tantamount to abandoning science for magic. Relinquishing the PSR might also undermine our self-conception as rational creatures in a world that is open to our understanding. As ongoing debates of this kind highlight, the examination of cosmological arguments implicates elemental questions about the nature of existence and how we come to understand it.

While critiques of the PSR have direct implications for discussion of God as a necessary being, a distinctive set of objections confronts claims that God is a contingent being. To say that God exists contingently is to say that God's existence did not have to be the case. It could have turned out that there was no God. Theists disagree over whether God's existence is necessary or contingent. For instance, while Leibniz aims to establish the existence of a necessarily existing God, we have seen that Swinburne makes the case for a contingently existing God. Swinburne argues that recognizing a necessary God has unacceptable implications. In particular, if a necessary being created the world, then everything in the world would also be necessary.[82]

79 See Mawson, *Belief in God*, 159.

80 Mawson, *Belief in God*, 159–60.

81 See Chapter Three. It remains controversial among scientists and philosophers alike whether we should say that a particular event occurs without a reason when it follows statistically predictable regularities considered in the aggregate.

82 Davies, *An Introduction to the Philosophy of Religion*, 62–63.

This is a problem if one believes, as Swinburne does, that the world includes contingent things, such as chance and free will.[83] Positing a contingent God, however, raises a difficult question for theists. If God's existence is unexplained, what reasons do we have for believing that God is the foundational fact that explains everything else?

One advantage of positing a necessary (rather than contingent) God is that it identifies a reason for viewing God as *uniquely* able to explain the world's existence. The argument is that only a necessarily existing God can explain a world of contingent things. The case for a contingent God cannot rest on the same reasoning. Consequently, assertions of a contingent God face this question: If God's existence is a brute fact, then why not view some other brute fact as the unexplained starting point of explanation?[84] In this vein, some thinkers prefer positing the existence of the universe itself as a brute fact.[85] This approach eliminates the need to bring an additional unexplained entity—God—into the picture.

Swinburne argues that God is a better explanation than the universe itself because it is simpler. The argument hinges on a number of assumptions, any of which might be challenged. For example, Swinburne assumes that simplicity is the central criterion for evaluating explanations, and that the simplest explanation is the one that posits the fewest entities. Further, he defines God as the simplest possible person, and argues that God possesses all the attributes of persons (including power, knowledge, and goodness) to an infinite degree, because infinite properties are simpler than properties manifested to any lesser degree. According to Swinburne, he only needs to posit one brute fact: the existence of God. If you grant him that assumption, everything else follows, including all the divine properties, and the fact that an infinitely good, powerful, and knowledgeable God could and would create the world. By contrast, materialism, Swinburne claims, must separately posit the existence and properties of every physical entity in the universe. Each fact about each distinct physical entity is independent of every other fact about every other entity. The nature of the physical world does not allow us to view facts about one object as entailed by facts about any other object. However, a potential line of objection to Swinburne's case for a theism would challenge his claim that infinite properties are simpler than those manifesting properties to a lesser degree. Without this assumption, the case for theism's simplicity falls apart. At the same time, one might argue, one must only posit a small set of natural laws—or even a single theory of

83 Ruling out free will would be especially troubling for a theist like Swinburne who places so much emphasis on the capacity of persons to act freely.

84 Mawson, *Belief in God*, 150.

85 Reichenbach, "Cosmological Argument."

everything[86]—to account for a great many truths about the universe. Thus, only a couple tweaks to Swinburne's assumptions are needed to yield the conclusion that the simplest explanation may be attained by positing only a small set of scientific facts, or simply the existence of the universe itself.

Having considered challenges to particular arguments for God's existence, we next consider objections to theism of a more general character. One broad challenge contends that the idea of God is incoherent or self-contradictory. If the description of a thing does not make sense, we normally assume that such a thing could not exist in reality. We know that no round square exists because the idea contains a contradiction. In that spirit, critics argue that God does not exist because the idea contains contradictions.

God is commonly conceived of as omnipotent, omniscient, and omnibenevolent. Opponents claim that each of these attributes is incoherent. A longstanding argument of this kind asserts that the idea of omnipotence leads to logical contradictions. The challenge is presented in the form of a question: Could God create a rock so heavy that God could not lift it? Objectors say that either a "yes" or "no" answer identifies something that God cannot do. A "yes" answer means that there is an object that God cannot lift. A "no" answer means that there is an object that God cannot create. Either way, there is something that God cannot do, which contradicts the meaning of omnipotence. Theists have a ready response: to say that God is omnipotent is not equivalent to saying that any question asking if God can do X has a "yes" answer. Rather, omnipotence is the ability to do anything that is logically possible. The existence of a rock so heavy that God cannot lift it falls outside the scope of logical possibility and, thus, poses no threat to divine omnipotence. Observing that God cannot perform an act described in a meaningless way does not reveal a limitation on God's power, just as it is no constraint on divine power that God cannot create a round square. A theist may allow that God cannot do nonsense without embarrassment or fear of inconsistency.[87]

Critics of theism also argue that certain combinations of attributes commonly ascribed to God cannot be held by the same being without contradiction. For example, some claim that omnipotence is incompatible with omniscience because a being with perfect knowledge of the future lacks the power to contravene that already known future.[88] Others insist that omnipotence is incompatible with omnibenevolence because a perfectly good being is not free to violate the requirements of acting for the good.[89] Theists

86 See Chapter Three.

87 Mawson, *Belief in God*, 28–29.

88 Richard Dawkins, *The God Delusion* (Mariner Books, 2008), 77–78.

89 Beverly Clack and Brian R. Clack, *The Philosophy of Religion: An Introduction*, 3rd ed. (Polity, 2019), 84.

have responses to these objections as well,[90] and the debate over questions of this kind is vast.[91]

We will concentrate in particular on charges of incoherence that challenge the combination of perfection and personhood in a single being. While theists have disagreed on virtually every aspect of God's nature,[92] many prominent theistic views have held both that God is maximally great and that God has critical features of personhood. A tension arises from the combination of two ideas: (1) that God is immanent in the world and like us in certain ways, and (2) that God transcends the world and is radically different from us.[93]

We have been focusing on theism in the broadest terms as the view that the explanation for the world's existence lies in the intentional actions of a personal being. In these accounts, God created the world through God's own agency as a being capable of willfully forming and acting according to intentions. In this conception, God is "a someone not a something."[94] While precisely what it means to speak of God as a person is controversial, God's personhood has often been associated with a variety of features, including rationality (the ability to reason) and intentionality (the ability to form and act according to intentions).[95] Thus, God has a mind not only in the sense of holding ideas, but also in the sense of possessing consciousness, beliefs, and understanding.

---

90 For instance, discussion often centers on whether God's unique relation to time might reconcile any apparent contradictions between omnipotence and omniscience, and whether conceiving of omnipotence as the power only to do what one wishes might reconcile any apparent contradictions between omnipotence and omnibenevolence. Mawson, *Belief in God*, 53–54.

91 For one example from this enormous literature, see generally J. Hoffman and G.D. Rosenkrantz, *The Divine Attributes* (Blackwell, 2002).

92 For example, theists have long debated the extent to which God transcends the world in which we live (rather than being immanent within it). Joseph W. Koterski, *An Introduction to Medieval Philosophy* (Wiley-Blackwell, 2008), 39–45. Some theists deny that the features of personhood properly apply to God at all, except in a metaphorical sense. Charles Taliaferro, "Personal," in *Philosophy: A Guide to the Subject of Religion*, ed. Brian Davies (Georgetown University Press, 1998), 96. Prominent theologians in each of the major Abrahamic traditions view God as a simple being—one that does not possess parts—with radical implications for how different God is from human beings (and all other creatures). If God has no parts, it has been argued, then it is not even correct strictly speaking to say that God has attributes. Instead, on this view, God is identical to the attributes, which are not really distinct from each other in the first place. The attributes are one, and God is one with the attributes.

93 See Koterski, *An Introduction to Medieval Philosophy*, 40–41.

94 Mawson, *Belief in God*, 13.

95 Some critics of theism deny that personal intention represents a distinct kind of explanation from the kind of explanations offered by science. In this view, what we speak of as a person's will or intentional agency is entirely explicable in terms of the same sorts of natural processes that are the subject of scientific theories. Clack and Clack, *The Philosophy of Religion*, 33. Dating to at least as early as the Presocratic philosopher Democritus, some thinkers have viewed persons as ultimately no less material in nature than any other perceptible objects in the world. Taliaferro, "Personal," 97. Such a view, if correct, makes personal explanation an illusion, thus nullifying theism as an explanation for the world's existence that rests on a genuinely distinctive basis. Clack and Clack, *The Philosophy of Religion*, 37–38. However, our focus will not be on this kind of objection but, rather, on arguments that God's personhood is in tension with other features often associated with God, particularly those related to divine perfection.

Divine personhood also has implications for God's ability to interact with human beings. God cares about the world and the people in it. Furthermore, God can communicate with people, and may react or respond to what particular individuals do.[96] Although theists disagree over how literally to interpret canonical texts,[97] many biblical narratives—such as the story of God's negotiating with Abraham over the fate of the Sodomites—suggest an interactive God who is capable of sustaining ongoing relationships with human beings.[98]

Nevertheless, God is a very different kind of being from ourselves. Unlike flawed human beings, God is commonly understood as a maximally great or perfect being.[99] We are inherently finite creatures, but God has the best qualities to the highest possible degree. Thus, God is not merely rational, but knows and understands all truth, and God is not merely good, but always does the best possible thing. In short, God is the most perfect person possible.[100]

Many find the union of divine perfection and personhood compelling. A personal God means that we can relate to—and identify with—the being who is responsible for the world's existence. Such a conception suggests that we can understand our own existence in terms of a relation with the very reason why we—and anything at all—are here in the first place. Theism also draws appeal from the idea of a maximally great being. A most wonderful God is not only a being with whom one may have a relationship, but one with whom it is extraordinarily rewarding to do so. It is intelligible to communicate with an interactive God, and a maximally great being is one who merits our attention and devotion.[101]

While the combination of personhood and maximal greatness is attractive, it also raises questions about theism's coherence. Potential difficulties concern the question of whether God can change. Central to these worries is the possibility that God's perfection implies immutability, thereby undermining divine personhood.

Although theists disagree over God's mutability,[102] many have seen perfection as ruling out the possibility of change. A most perfect being, the argument goes, would have no reason to change. If a being is already maximally great, then any change can only make it worse, introducing deficiency and imperfection. Yet immutability is arguably at odds with personhood. We typically think of a person as someone who engages in intentional actions,

96 Mawson, *Belief in God*, 13–18.

97 Davies, *An Introduction to the Philosophy of Religion*, 182–83.

98 Genesis 18:22–33; Taliaferro, "Personal," 96.

99 We saw earlier in discussing ontological arguments that Anselm defines God as the greatest conceivable being, while Descartes defines God as the most perfect being, but theistic conceptions of this kind are not limited to the context of those advancing ontological arguments for God's existence.

100 Mawson, *Belief in God*, 18–19.

101 Taliaferro, "Personal," 95.

102 Mawson, *Belief in God*, 37.

which seems to entail change. A being who has formed an intention to do something is at least in that respect different from one who has not formed that intention. Similarly, a being who has engaged in action A is at least in that respect different from one who has not engaged in action A. Taking an action adds a new fact to an individual's narrative.[103] Persons not only act but also have experiences that alter their history.[104] In addition, persons acquire knowledge. God is commonly thought of as knowing everything that it is possible to know. And a being who learns is one who does not know something at one time but comes to know it at another, which suggests undergoing a kind of change.[105] More generally, due to the various ways that persons change through intentional actions, experiences, learning, and so on, we think it fitting to speak of them as developing. Persons do not remain exactly the same.

Another potential difficulty concerns God's motivations for acting. Theism as an ultimate explanation posits that the world is a product of divine intention, but we may wonder whether a perfect and immutable God could have a reason for engaging in the action of creating the world.[106] What could be the impetus for an already flawless and maximally excellent being to effect change?

Some have thought that God's perfection implies not only immutability but also invulnerability. According to this view, God cannot be affected by anything external, which implies that God cannot feel anything in response to other beings. That would mean that a perfect God could not love, feel empathy, or suffer on behalf of any other person or creature.[107] It would also suggest a lack of divine responsiveness to human beings.[108] The implications for the relation between God and the world are far-reaching. A God who does not act on or respond to the world cannot be part of an ongoing, dynamic relationship with human beings. For persons to share a joint narrative, each of the individual parties must be capable of participating in a narrative. Yet a God who does not change or react to human beings seems ineligible for participation in any kind of dramatic arc or joint story.

We cannot consider here in detail the wide-ranging and longstanding philosophical and theological debates implicated by whether divine perfection is consistent with divine personhood. In the broadest terms, theistic responses to the charge that the idea of a perfect person is self-contradictory may take three principal forms: denying God's perfection, denying God's personhood, or showing that divine perfection and personhood

103 Davies, *An Introduction to the Philosophy of Religion*, 169–70.
104 Davies, *An Introduction to the Philosophy of Religion*, 175.
105 Davies, *An Introduction to the Philosophy of Religion*, 171.
106 Davies, *An Introduction to the Philosophy of Religion*, 176.
107 Davies, *An Introduction to the Philosophy of Religion*, 175.
108 Davies, *An Introduction to the Philosophy of Religion*, 169.

are compatible.[109] The challenges posed by the objection reflect a tension regarding the distance between God and humanity. Denying God's perfection (or substantially downgrading its meaning) makes it easier to highlight the closeness between people and God, but undermines God as a being of infinite greatness. Denying God's personhood (or substantially downgrading its meaning) makes it easier to preserve God's unlimited greatness, but undermines God as *someone* with whom people may have a relationship. It would be hard for many theists to embrace a view precluding God from loving the divine creation, or caring about people's prayers, much less responding to them.[110]

The next general objection that we will consider—based in the "problem of evil"—charges that theism is implausible in light of the bad things that we observe in the world. Theism maintains that an all-good God created the world, and the objection argues that the world would be a better place if that were true. While there are many bad things in the world, critics of theism have focused above all on pain and suffering. We would expect God to create the best possible world, and yet we can imagine worlds that are substantially better than this one.[111]

---

109 One prominent strategy that theists have used to address apparent contradictions in the idea of God concerns the relation of God to time. The core idea—most famously associated with the early medieval philosopher Boethius—is that God does not exist within flowing time in the way that we do. In this view, God is eternal, not in the sense that God has always existed and continues to flow through time forever, but, rather, in the sense that all of reality is always present to God. This understanding dissolves certain apparent contradictions in the idea of God, because many such apparent contradictions are premised on the notion that God moves through time. For example, this conception of God's eternality may help to reconcile divine omnipotence with divine freedom and omniscience, because God's infinite knowledge need not entail that God knows things *before* they occur. It may also seem to address potential difficulties concerning the combination of divine perfection and divine personhood, since many of them, too, are based in the premise that God's relation to time is similar to our own. We have seen that many of these difficulties concern the potential tension between immutability and personhood. We noted, for example, that acting involves not having an intention at one moment but having it at another. However, if God does not exist within flowing time, then we cannot speak of God as being different *at one time* versus *another time*. Thus, the apparent contradiction may not be genuine. See, generally, Paul Helm, "Eternality," in *Philosophy: A Guide to the Subject of Religion*, ed. Brian Davies (Georgetown University Press, 1998), 75–79.

110 See Michael J. Dodds, *The Unchanging God of Love*, 2nd ed. (Catholic University Press, 2011), 1–2; Mawson, *Belief in God*, 12, 38; Taliaferro, "Personal," 99.

111 Our discussion concentrates on the "evidential problem of evil," which maintains that theism is highly unlikely given the extent of the world's evils. See Brian Davies, "The Problem of Evil," in *Philosophy of Religion: A Guide to the Subject*, ed. Brian Davies (Georgetown University Press, 2007), 163–64. Another version—known as the "logical problem of evil"—claims that the existence of evil makes theism not just implausible, but logically impossible. Davies, *An Introduction to the Philosophy of Religion*, 247. J.L. Mackie, for example, advanced a well-known argument of this kind. "Evil and Omnipotence," *Mind* 64 (1955): 26. The logical problem of evil claims that the existence of any evil at all represents a self-contradiction within theism. Consequently, the logical problem of evil is vulnerable to a demonstration that any amount of evil might be justified by the greater goods that it makes possible. Many critics of theism stress the evidential problem of evil because it is not necessarily defeated by a showing that a single instance of evil—no matter how small—might be compatible with theism. Mackie himself later shifted his argument against theism from a logical to an evidential version of the problem of evil, maintaining that it does not present a conclusive proof, but gives rise to a strong presumption against God's existence. J.L. Mackie, *The Miracle of Theism* (Oxford University Press, 1982), 176.

Among the objections to theism of this kind, none are better known than those offered by the eighteenth-century philosopher David Hume.[112] Stressing the pervasiveness of "human misery,"[113] Hume wrote that each person's "first entrance into life gives anguish to the new-born infant and to its wretched parent: Weakness, impotence, distress, attend each stage of that life: And it is at last finished in agony and horror."[114] More specifically, Hume provided a litany of horrors that afflict people's lives, including diseases, poverty, oppression, injustice, fear, and anxiety.[115] Nor is suffering limited to human beings, since "every animal is surrounded with enemies, which incessantly seek his misery and destruction."[116]

The last chapter discussed the problem of evil as an objection to axiological theories. The thrust of the argument is similar when the problem of evil is the basis of an objection to theism. As an argument against axiological theories, the objection from evil charges that the world does not look like it is based in and directed toward goodness. As an argument against theism, the objection charges that the world does not look like it was created by an all-good God. In both contexts, the objection is based in an alleged contrast between the actual world and a world supposedly sourced in goodness, whether the product of an all-good God or goodness itself. Consequently, many of the prominent arguments have similar force in the two contexts.

---

112 We first met Hume in discussing questions about whether the natural laws of science are rooted in necessity. See Chapter Three.

113 Hume, *Dialogues concerning Natural Religion*, 193.

114 Hume, *Dialogues concerning Natural Religion*, 194. Hume presents his arguments in the form of a dialogue among three characters: Cleanthes, Demea, and Philo. There is some uncertainty and disagreement regarding the extent to which any character functions as a mouthpiece for Hume's own ideas. Clack and Clack, *The Philosophy of Religion*, 40 n. 92. Nevertheless, it is commonly accepted that many of the views expressed by Philo represent Hume's own opinions. See Norman Kemp Smith, "Introduction," to Hume's *Dialogues concerning Natural Religion* (Thomas Nelson and Sons, 1947), 59. At any rate, regardless of Hume's personal views, the arguments he presented have been influential, and they have also been expressed by many other thinkers.

115 Hume, *Dialogues concerning Natural Religion*, 193–96.

116 Hume, *Dialogues concerning Natural Religion*, 195. Philo is especially interested in challenging theistic arguments which seek to show that the good things in the world—such as life evincing tremendous complexity—provide strong evidence for God's existence. In opposition to design arguments of this kind, for example, Philo argues that the nature of the world does not warrant an inference by analogy that it is the product of God's design. If anything, he says, the evidence is more consistent with a world that has grown like an organism. Hume, *Dialogues concerning Natural Religion*, 179–80. In further attacking the way that thinkers like his interlocutor Cleanthes argue from the world's character to a perfect God, Philo contends that various "inexplicable difficulties in the works of nature" suggest instead that the world is "only the first rude essay of some infant Deity, who afterwards abandoned it, ashamed of his lame performance" or that "it is the work only of some dependent, inferior Deity." Hume, *Dialogues concerning Natural Religion*, 166, 169. At other times, Philo relies on the existence of tremendous evil to argue not only that the design argument is flawed, but that theism, more generally, is implausible. In either case, the gist of the argument is similar: the nature of the world makes it unreasonable to suppose that it is the product of a perfect being. Hume, *Dialogues concerning Natural Religion*, 211.

Nevertheless, theism's focus on the attributes and intentions of a personal being naturally have an impact on the discourse.[117] The argument is commonly stated in terms of why an all-knowing, all-good, all-powerful God would allow evil in the world, given that such a being would know about evil, wish to eliminate it, and have the power to do so. As Hume concisely framed the issue, "Is [God] willing to prevent evil, but not able? Then is he impotent. Is he able, but not willing? Then is he malevolent. Is he both able and willing? Whence then is evil?"[118]

The objection from evil has generated an enormous amount of debate, and theists have offered a wide variety of responses in defense of God's existence.[119] We will concentrate on a particularly influential kind of defense which claims that the world's evils are justified by the existence of goods that are inextricably tied to them.[120] These goods are so great that they outweigh the evils that we see in the world. Since the evils are inseparable from greater goods, the world would not be better off without them. The complex relation between goods and evils means that we cannot assess the world's overall goodness in a piecemeal fashion. A world with nothing but all of the good things that we can imagine is not possible. We must take a more holistic view. This greater goods argument is often combined with the claim that human beings' limited perspective makes it difficult for us to appreciate all of the ways that apparently evil things are part of an overall whole that is wonderfully good. As one scholar writes: "If we had access to God's whole plan ... we might be able to appreciate that a given region of human

117 As an argument against theism, the objection from evil dates to well before the advent of Christianity. The Greek philosopher Epicurus (341–270 BCE) discussed the issue, for example. Clack and Clack, *The Philosophy of Religion*, 94. It has also long been a subject of intense interest among theologians, including the early medieval bishop Augustine, who was preoccupied with it throughout his career. Kaye, *Medieval Philosophy*, 53–61.

118 Hume, *Dialogues concerning Natural Religion*, 198.

119 For example, some prominent theologians—Augustine and Aquinas most famous among them—deny that God is responsible for evil on the grounds that evil is not a distinctly existing thing with its own positive qualities. According to this view, evil refers merely to the absence of a good—where a good thing might be—rather than to an independently existing thing in itself. Davies, "The Problem of Evil," 169, 195. Another argument defends against the charge that a perfect God would create the best possible world by claiming that there is no uniquely determinative best possible world. Just as there is no such thing as a highest number—because one could always add greater quantity to any number no matter how large—there is no best possible world, because one could always add more good things to make a world better. Michael Tooley, "The Problem of Evil," in *The Stanford Encyclopedia of Philosophy*, ed. Edward N. Zalta (2021), § 5.2, accessed December 10, 2024, https://plato.stanford.edu/archives/win2021/entries/evil/. As one scholar writes: "God's inability to create the greatest state of happiness is seen to be no different from his inability to create the greatest integer. Neither diminish His might, if we agree that He need not accomplish what is logically impossible." G. Schlesinger, "The Problem of Evil and the Problem of Suffering," *American Philosophical Quarterly* 1, no. 3 (1964): 245.

120 In itself, the idea that goods are linked with evils is commonplace. Unpleasant medical treatments make us healthier. Difficult moments while working on a project add to the sense of accomplishment felt upon completion. And we may appreciate children's acquisition of new abilities all the more in light of the challenges faced along the way.

experience, which, though in isolation looks ugly, viewed as a part of the entire pattern, enhances the over-all beauty and goodness of the whole."[121]

Advocates of the greater goods defense have laid special emphasis on the good of free will. The ability to make genuine choices in itself is a good commonly cherished, as reflected in the importance that we place on autonomous decision making in our own lives. Freedom also makes possible many great things, such as feeling pride in completing a project, doing the right thing despite other possibilities, and developing relationships of mutual caring with chosen partners. Freedom is necessary for these goods, which are more valuable than if the outcomes were imposed upon us against our will. However, the existence of freedom inevitably brings evils in its wake. One cannot freely choose to act well unless one also has the option to choose poorly.[122] People having that freedom means that some will make errors or intentionally do evil.

If some goods inevitably bring evils in their trail, others evince a yet deeper interrelation with evils. The realization of some goods depends on the existence of evils. Goods of this kind include the instances in which people exercise virtues, such as perseverance, forgiveness, and compassion. One can only persevere in the face of adversity, for example, and acts of forgiveness involve a response to perceived wrongs. Compassion, too, depends on awareness of difficulties experienced by others. As Swinburne writes, "it is good to have a deep concern for others; and the concern can be a deep and serious one only if things are bad with the sufferer."[123]

Some theists have emphasized that there is special value in the development of a good character. Referring to this as the good of soul-making, John Hick writes that when a person builds an admirable character over time, "the individual's goodness has within it the strength of temptations overcome, a stability based upon an accumulation of right choices, and a positive and responsible character that comes from the investment of costly personal effort."[124] This good is inseparable from the existence of evils, because a good character is more valuable when it is the product of free choices than if we lacked freedom or came into the world already gifted with virtuous souls.

The goods associated with soul-making help to address a particularly difficult challenge for theists: explaining the existence of natural evils. Unlike moral evils—the evils resulting from individuals' free choices—natural evils cause pain and suffering but are not traceable to human actions.

121 Schlesinger, "The Problem of Evil and the Problem of Suffering," 245. Augustine is one of many theologians who have made similar points in discussing the existence of things in the world that in themselves appear not to be good. See Augustine, *City of God*, 11.23.

122 Davies, *An Introduction to the Philosophy of Religion*, 262–63.

123 Swinburne, *The Existence of God*, 240.

124 Hick, *Evil and the God of Love*, 255–56.

Examples include earthquakes, droughts, and diseases. Even if the goods tied to free will explain the existence of moral evils, they do not explain why God would allow the destruction caused by natural disasters. In response, theists contend that natural evils provide us with knowledge that is essential to our moral freedom. We must understand the goods and evils that we can do *before* making genuine choices.[125] Moreover, natural evils are indispensable to our living in an environment with sufficient opportunities for learning to do the right things and expressing the appropriate emotional responses in the face of difficulties.[126]

Theists also justify the existence of natural evils on the grounds that they are an inevitable outgrowth of natural laws. As the fine-tuning problem highlights,[127] our lives as creatures in the world depend on the operation of consistent regularities in the ways that physical things behave. The reliable applicability of natural laws is critical to the stability of the matrix within which we conduct and plan our lives over time. Like free will, though, the natural laws inevitably bring evils with them, since their consistent workings entail that they will sometimes lead to events that in themselves seem anything but good.

The greater goods defense aims to exonerate God by showing that the world's evils—when understood properly within their larger context—make the world the best that it can be. Eliminating the existence of evils would not improve the world, because they are inextricably linked with greater goods that more than make up for them. Critics respond by identifying harms that do not seem necessary for the existence of greater goods. Even assuming that an all-powerful, all-knowing, all-good God would allow some evils for the achievement of greater goods, we would expect such a God not to allow any more pain and suffering than was absolutely necessary for the attainment of greater goods. Evil may be consistent with God's existence, but gratuitous evil is not.

To make the case that the world includes gratuitous evils, critics contend that God could have limited the scope of free will so that people could not perpetrate the most destructive kinds of harms, such as genocide, torture, and other atrocities. God also could have designed the natural laws—and the universe's initial conditions—so that human beings would experience less pain and fewer natural disasters would occur. In addition, God could intervene to prevent the worst instances of both moral and natural evils.[128]

Critics have also contended that people could exercise virtues by helping others to achieve great things in the face of challenges—as in athletics,

125 Davies, *An Introduction to the Philosophy of Religion*, 266.

126 Swinburne, *The Existence of God*, 240.

127 See Chapter Five.

128 Schlesinger, "The Problem of Evil and the Problem of Suffering," 245.

art, or intellectual pursuits—rather than by helping them through awful pain and suffering.[129] Even if significant harms are indispensable to soul-making, there is much pain and suffering that seems detached from the goods of soul-making. For example, people often suffer great pain in their last moments of life when no opportunities remain for enhancements to their character. Moreover, some suffering is so relentless, overwhelming, and soul-*breaking* that talk of soul-making seems out of place.[130] In addition, the terrible pain and agonizing deaths of countless animals is built into the ecosystem yet cannot be tied to character building.[131]

One response to the greater goods defense, then, focuses on the *quantity* of pain and suffering, contending that a large amount could be eliminated without diminishing the overall good in the world. Another response focuses on the *distribution* of pain and suffering. We can imagine a world in which the overall good for each individual—taken one at a time—outweighs and justifies the evils endured by that particular person. But that is not our world. There are numerous infants and children subjected to terrible evils who die at a young age with no chance to experience greater goods that could make up for them. The objection charges that an all-powerful, all-knowing, and all-good God would not allow even a single instance of such unredeemed misery, where the greater goods do not outweigh the evils for that individual.

One theistic reply to objections based in the quantity of evil insists that every quantum of evil contributes in some way to greater goods. As Swinburne writes: "Each small addition to the number of actual or possible bad states makes a small addition to the number of actual or possible good states." A world with substantially fewer evils, he continues, "would be a toy-world; a world where things matter, but not very much; where we can choose and our choices can make a small difference, but the real choices remain God's."[132] To buttress this kind of argument, some theists appeal to greater goods that we cannot fully appreciate from our limited perspective. The world's evils are essential components of a larger framework of goods that is—at least for the time being—not transparent to us.[133] In this vein, both Hick and Swinburne concede that only goods forthcoming in another life can ultimately justify the worst evils that we encounter in this world. Regarding the distribution of evils, Swinburne allows that "God would not have the right to give anyone an earthly life that is on balance bad unless

129 G. Stanley Kane, "The Failure of Soul-Making Theodicy," *International Journal for Philosophy of Religion* 6 (1975): 2–3.
130 Clack and Clack, *The Philosophy of Religion*, 98.
131 Hick, *Evil and the God of Love*, 309–17.
132 Swinburne, *The Existence of God*, 264.
133 Hick, *Evil and the God of Love*, 337–41.

he provided for them a compensatory period of good life after death."[134] For some, then, faith in hidden goods provides a compelling antidote for the problem of evil. At the same time, others find this sort of response unsatisfying, seeing it as merely gesturing toward an unresolved mystery rather than offering a comprehensible solution.[135]

Living in the Divine Universe

In the last chapter, we noted ways in which the good universe differs from both the scientific and mathematical universes. Though for different reasons, neither science nor mathematics recognizes anything that is special about human beings. While we experience our lives as revolving around value-laden notions like the pursuit of joy and meaning, these are in no way central to the scientific or mathematical universes. By contrast, axiological theories describe a universe in which the very foundation of existence—goodness itself—resonates with our concerns. From an axiological standpoint, the cosmic narrative may even have been directed from the beginning toward our own existence. In these respects, the divine universe shares a great deal with the good universe. In theistic accounts—as in axiological ones—goodness exists independently of anything in the physical world. The universe's existence is an objectively good thing, and we may even occupy a special place within an ongoing cosmic narrative.

Nevertheless, despite the substantial similarities, the divine universe also differs from the good universe in fundamental ways. While axiological accounts place goodness at the foundation of reality, and recognize it as giving rise to conscious creatures like ourselves, goodness does not have its own consciousness. The cosmic narrative in the good universe might have been directed toward our existence, but the good is not itself aware of us. For Plato—the seminal and quintessential axiological thinker—the form of the good is an impersonal being. Individuals might yearn for and love the good, but the good does not return the favor.

In the divine universe, more than in any other, the world's ultimate explanation shares features that are defining of our own existence. Like us, God is a person who makes things happen through intentional actions. God

134 Swinburne, *The Existence of God*, 264.

135 Some individual religions appeal to faith-based narratives in which the suffering in this world is only a prelude to an incomprehensibly greater, infinite reward to be enjoyed in another life. Philosopher John Hick, for example, writes that a proper response to the problem of evil "must be eschatological in its ultimate bearings ... instead of looking to the past ... it looks to ... that ultimate future to which only faith can look.... [We] must find the meaning of evil in the part that it is made to play in ... the kingdom which is yet to come in its full glory and permanence." Hick, *Evil and the God of Love*, 261.

is not some thing, but *someone*. The divine universe is one in which the very foundation of reality can love us back. Our relationship to the reason for the world's existence can be just that: a *relationship*, at least in some ways similar to what we have in mind when speaking of our connections with other people. If there is an eternal being who knows and understands us, we can never be absolutely alone.

The good and divine also have both similarities and differences with respect to the status of truths about value. Like the good universe, the divine universe has objective truths about ethics, what makes for a worthwhile life, and other questions of value. While theists have long debated the relation between divinity and goodness, they have commonly assumed that God is perfectly good and created the world because it was good to do so. Thus, as in the good universe, the divine universe has standards of goodness that are fully independent of our own opinions. We need not think that talk about morality and other value-laden ideas are just the product of our own creative imaginations. Rather, it is possible to lead one's life in accordance with guidelines that have a real existence apart from our particular desires and inclinations.

Nevertheless, the place of the good in theistic accounts is not entirely the same as in axiological accounts. Indeed, one source of theism's appeal is the manner in which it incorporates advantages enjoyed by axiological theories while avoiding some of their potential difficulties. Let us recall some of an axiological approach's appealing features. For one, grounding reality in goodness can help us make sense of the pervasive role that value plays in our lives. In addition, faced with the fine-tuning problem, axiological theories offer an explanation for the universe's hospitality to intelligent life. The tight relation between being and goodness also has implications regarding our place in the world that many find congenial. If goodness is the explanation for the world's existence, then our cosmic home is benevolent and we have reasons to hope for a better future. Our own existence may even be an end toward which the cosmos was directed from the outset.

At the same time, an axiological approach requires beliefs that may be hard for some to take on board, particularly the idea that goodness has a reality that is independent of anything else. From an axiological standpoint, if there were no world, no people, and no beings of any kind, there would still be truths about value. However, some observers reject the idea that truths about value could have a real existence in the absence of anyone for whom particular things actually hold value. Resistance to the idea of free-standing value is closely related to the view that value is only intelligible when understood in terms of the interests of living beings. As a result, it is easier for some people to accept an account rooted in a (divine) person who is motivated by goodness than an account that identifies goodness itself as the ultimate explanation. Value and intentional action may even seem so intertwined that they only make sense in combination. This is especially so if one supposes—as many do—that intentional actions are by their nature guided by value considerations and that value only takes on meaning due to its implications for particular actors.[136]

Given what many perceive as the implausibility of freestanding value, a potential advantage of theism is that it retains a vital place for goodness while uniting it with personhood. In many theistic belief systems, as in the Abrahamic tradition, God not only has infinite power but is also perfectly good. Since God's goodness motivated the divine act of creation, theists may conceive of the world as imbued with purpose and directed toward worthy ends without having to posit the existence of freestanding truths about value. In a theistic account, reality's foundation lies in the intentional, deliberate actions of a deity. If that deity is perfectly good, then both personhood and goodness underlie the world's existence. As a supernatural person of unlimited power, God chooses to create the world. Thus, we can understand our cosmic home as the product of an omnibenevolent being's intentional acts.

Of course, theism entails much more than a conjoining of personhood and goodness. It also centers on an extraordinary kind of person. Human beings are part of the world for which we seek an explanation, and our creative capacities do not encompass the ability to produce material existence from nothing. Thus, while holding onto the idea of personhood—entailing consciousness, thought, and intentional agency—theism attaches it to a being who differs from people in fundamental ways. In contrast with common assumptions about human beings, God is an immaterial divine being capable of acting through intentional actions without the mediating movements of a physical body. Moreover, the scope of what God can achieve greatly exceeds anything that is humanly possible. God's extraordinary power includes the ability to produce physical existence from nothing.

When it comes to our ability to grasp the most fundamental truths, theism suggests a curious combination of optimistic and pessimistic implications. This is due largely to the way in which it responds to two competing impulses. One impulse pushes in the direction of making possible a meaningful connection with God. This yearning is fulfilled by a God who is more like us, a God who has characteristics we associate with interpersonal relationships. A God with whom we can interact is one who can engage with us as a character in an ongoing narrative.

Similarities of this kind between us and God point to ways in which theism as an ultimate explanation is remarkably accessible. The idea of a

136 Armour, "Values, God, and the Problem about Why There Is Anything at All," 155–56.

person acting on an intention could not be more familiar. We intuitively grasp the notion of pursuing something one deems to be worthwhile, as this kind of purposiveness in seeking good things is integral to our lives. Moreover, theism opens the door to direct communication with the very source of being. We might speak *metaphorically* of scientific natural laws, mathematical truths, or moral principles as revealing themselves to us, but only in the divine universe can the world's foundation *literally* convey fundamental truths to us. If God is a person capable of thought and communication who understands us and loves us, then there is the possibility that God would choose to further our understanding through the written or spoken word. Even without verbal revelation, God might have formed our minds with the intention of facilitating our ability to gain vital knowledge.

At the same time, a God who was too much like us could not be the source of all existence. Thus, a competing impulse pushes in the direction of a God whom we can plausibly recognize as the origin of all being. This longing can only be satisfied by a God who differs from us in the most deep-seated ways, a God with capacities that far exceed the scope of our imaginations. God is unlimited, perfect, and eternal, all characteristics that we cannot fully grasp. While God is in some ways close to us—allowing for accessibility—in other ways God transcends our capacities, which renders the foundation of reality profoundly incomprehensible. By placing a person at the foundation, theism suggests that our nature is deeply in tune with the cosmos, while also conceiving of God as a person who is extraordinarily removed from us.

Religious traditions have long grappled with the problem of how finite creatures like us could interact with an unlimited being. One common way to address this tension is by recognizing special individuals with whom God chooses to communicate directly. Christianity's worship of a person who is at once fully human and fully divine is a particularly clear instance of a theology that centers on the tension between the immanent and the transcendent.

For a worldview based in a divine *person*, it may be only fitting to suppose that God has depths beyond our understanding. In a sense, this recognition of the limitations on our understanding of a person resonates with our daily lives. Even the people closest to us always remain beyond our complete understanding. It is in the nature of our existence as individual consciousnesses that we can never actually experience the world from another's vantage point, no matter how much we may travel a similar journey.

Perhaps that unbridgeable gap between one mind and another is a clue to the reason that the idea of faith figures so prominently both in everyday relationships and theism. Even in the absence of logical demonstrations that could provide definitive truths, we may commit to trusting certain things about another person. Based on observations we make in our relationships, we sometimes choose to believe in another's good intentions and favorable disposition toward us. There may be an analogy here with people's relation to the divine, even though the gap between us and God is so much greater. Faith in the theistic context may connote a commitment to live *as if* certain things were true about another (divine) person, even though it is inherently beyond our capacities to confirm those things in a definitive manner.

We noted earlier that some object to theism on the grounds of incoherence, charging that the personhood and perfection commonly associated with God cannot be reconciled. People naturally disagree about whether this kind of objection is decisive. What we can say with greater assurance is that the tension between God's immanence and transcendence is a distinctive element of theism. Life in the divine universe offers the prospect of deep interconnection with the very reason for our existence, while also suggesting that complete understanding—at least in this life—must always elude our grasp.

Questions for reflection and discussion

1. Is it possible to establish the actual existence of an object through reasoning alone?
2. Can we form a coherent idea of the greatest conceivable being?
3. What is the most compelling aspect of Aquinas's first-mover argument for God's existence?
4. Do you agree with Leibniz's assumption that there is a reason for everything?
5. How persuasive is Swinburne's argument that the ultimate explanation must be a brute fact?
6. Can we initiate events in a way that is independent of purely scientific mechanisms?
7. Which explanation of the world's existence is simpler: theism or science?
8. Could a perfect being act intentionally, care about other beings, and change over time?
9. What is the best argument that God's existence is consistent with the reality of suffering?
10. What is the most attractive feature of theism as an explanation for the world's existence?

7. The Lawful Universe

Nomological Explanation

Each of the approaches we have examined thus far—scientific, mathematical, axiological, and theistic—is rooted in a type of explanation that is familiar to us in our daily lives. We are accustomed to thinking of why-questions as having answers that speak in terms of physical properties, mathematical relationships, the value that different things hold for us, or a person's intentional actions. Additionally, each of these approaches is tied to a longstanding intellectual tradition that draws on a rich corpus of well-known thinkers and texts. This combination of familiarity and historical pedigree imbues the types of explanation that we have discussed with a certain facial plausibility as bases of explanation generally, and makes them intriguing candidates for answering the puzzle of existence.

While there are compelling reasons for taking scientific, mathematical, axiological, and theistic approaches to ultimate explanation seriously, we have also seen that each faces its own distinctive challenges. Notwithstanding the challenges, advocates of each of these approaches maintain that it remains promising as a path to explaining why the world exists. Others may consider the difficulties insurmountable and conclude that we should abandon the endeavor entirely. In this chapter, we consider another alternative: that the explanations we have examined are inapt for the task, but that another form of explanation can unlock a satisfying answer.

In contrast with advocates of the approaches previously examined, the thinkers discussed in this chapter do not present their views as falling within a longstanding tradition. Nor do they identify a canon of great texts as precursors to their own theories. Nevertheless, these accounts share enough in common that it is helpful to recognize them as instances of a distinctive strategy for explaining the world's existence. We can refer to this fifth approach as "nomological" after the Greek term *nomos*, meaning "law." In the most concise terms, a nomological approach is one that grounds the physical world's existence in an explanatory law that exists independently of the physical world. These laws are not physical in character, yet they bear ultimate responsibility for physical existence. Nomological thinkers turn away from the approaches that we have previously examined but not from the project of explaining the world's existence. As one nomological thinker writes: "If there is some explanation of the whole of reality, we should not expect this explanation to fit neatly into some familiar category."[1] Even if science, math, value, and theism are inapt for solving the puzzle of existence, that does not necessarily mean that nothing else can do the job.

Although the theories that we will discuss are based in different laws, the commonalities in their arguments shed light on the motivation underlying their adoption of a nomological approach. Nomological accounts generally assume that the world is rationally ordered. Most relevantly, this means that the most basic thing about the world—the reason for its existence—has an explanation that is open to our understanding. Consequently, nomological thinkers view the pursuit of an explanation for the world's existence as a worthwhile endeavor. Moreover, these thinkers purport to gain insights and make progress toward an answer by reflecting on the requirements of rationality. We can rule out some theories on the grounds that they do not fulfill the requirements for a satisfying explanation. Accounts grounded in scientific theories or mechanisms of physical causation, for example, are unavailing because they build in physical existence from the start rather than identifying a reason for it.

More generally, nomological accounts express frustration with the more familiar approaches to explaining the world's existence, which they regard as internally inconsistent or highly implausible. Thus, while supposing that there is a rationally comprehensible reason for the world's existence, nomological thinkers find that other major approaches are inadequate for the job. Declining either to embrace one of the familiar types of explanation or to abandon the quest, nomological thinkers propose independent explanatory laws that bear responsibility for the world's existence. Nomological thinkers do not try to fit these laws within the theories of existence that we have examined. Instead, they present their foundational laws as freestanding or

1 Derek Parfit, "Why Anything? Why This? Part 1," *London Review of Books* 20, no. 2 (1998): 26.

treat the quest for the world's existence as sui generis (representing a unique explanatory context). As we shall see, for example, Derek Parfit proposes a principle requiring the world's existence due to a special feature that it possesses, and Quentin Smith's foundational law holds that the simplest possible thing comes into existence in the simplest possible way.

The term "laws" comes with potentially misleading baggage, because we speak about them in so many contexts, from enacted legislation to scientific theories. Thus, it is important to clarify the sense in which we are referring to the idea of law here. In the context of nomological theories, laws have two salient features. First, they are efficacious, meaning that they bear direct responsibility for a state of affairs being the way that it is. These laws are not merely guidelines, standards, or aspirations with which people might or might not choose to comply. Nor are they merely convenient rhetorical devices for summarizing regularities or events that have their actual source elsewhere. Rather, these explanatory laws make things the case in the world that would not otherwise be so. Second, the laws that we are considering in nomological accounts enjoy an independent reality. They do not depend on the physical world or pre-existing things for their own existence. To explore how one might account for the world through a nomological approach, we will discuss the theories offered by philosophers Derek Parfit (1942–2017), Peter van Inwagen (b. 1942), Quentin Smith (1952–2020), and Robert Nozick (1938–2002).

Parfit's "Selector"

The assumption that the world is rationally comprehensible plays a pronounced role in Derek Parfit's work. Aiming to make reality more intelligible, Parfit prefers theories that leave as little unexplained as possible. All things being equal, it counts in favor of a theory if it purports to explain more than rival theories do. Indeed, Parfit's preference for explaining as much as possible underlies his interest in the quest for a reason why the world exists. Parfit is further motivated by Leibniz's claim that nothingness—the complete absence of a physical world—would have been the simplest and easiest state of affairs. Leibniz asserts this view in urging us to feel surprised by the fact that anything exists at all. If it would have been easier and simpler for nothing to exist, Leibniz asks, then why is there something rather than nothing?[2] Embracing Leibniz's reasoning, Parfit, too, thinks that the world's existence cries out for explanation. Citing Leibniz, Parfit writes

2 See Chapter Six.

that the fact that a universe exists "can take one's breath away" and provoke feelings of awe.[3]

At the same time, Parfit rejects well-known approaches to explaining the world's existence. He rules out scientific answers to the puzzle of existence, for example, as inherently circular. Earlier, we discussed the theoretical physicist Edward Tryon's proposal that the universe originated as a fluctuation in the quantum vacuum.[4] Parfit denies that rooting the universe in a quantum vacuum provides a satisfying answer to the puzzle of existence because the quantum vacuum is itself a physical thing.[5] Parfit also rules out axiological and theistic answers because neither can be squared with the fact that the world contains so many bad things, like pain and suffering. Thus, Parfit endorses the objection from evil,[6] which holds that a world based in goodness itself or the intentional actions of a perfectly good God would be better than the actual world.[7]

On what basis, then, does Parfit seek to identify a more promising approach to ultimate explanation? Interestingly, Parfit develops his alternative theory by extracting the general form of an approach that he otherwise dismisses. That is, while Parfit rejects any goodness-based explanation for the world's existence, he nevertheless finds something worthwhile in the basic structure of an axiological theory. At its core, Parfit observes, an axiological theory of existence encompasses two claims. First, an axiological theory claims that the world embodies a special feature—that it is the best of all possible worlds—and, second, it identifies that special feature as the reason for the world's existence.[8]

The general structure that Parfit deems promising is that the world exhibits a special feature—the "Selector"—which explains the world's existence. In this kind of account, the ultimate explanation is found in an explanatory law. That law makes it the case that the world exists—and exists in the form it does—because it possesses a special feature.[9] There is no mediating process, mechanism, or entity standing between the feature and the explanation for the world's existence. It is not that the feature provides a reason that motivates God to create the world, for example. The Selector itself is an explanatory law holding that such a world exists because it has this feature. Parfit explicates the idea as follows: "Of the countless ... ways that reality

3 Parfit, "Why Anything? Why This? Part 1," 25.

4 See Chapter Three.

5 Parfit, "Why Anything? Why This? Part 1," 24.

6 The problem of evil as an objection to axiological theories is discussed in Chapter Five, and as an objection to theistic explanations of the world's existence in Chapter Six.

7 Parfit, "Why Anything? Why This? Part 1," 26.

8 Parfit, "Why Anything? Why This? Part 1," 26–27.

9 Parfit, "Why Anything? Why This? Part 2," *London Review of Books* 20, no. 3 (1998): 22.

might be, a few have very special features. If such a possibility obtained, that might be no coincidence. Reality might be this way because this way had this feature."[10] Thus, any proposal of a particular Selector encompasses two distinct claims: (1) that the world embodies a particular special feature (which we may call the "descriptive claim"), and (2) that this special feature explains why the world exists as it does (the "explanatory claim").

The first part of arriving at a Selector is identifying a feature that the world has. Regarding this descriptive claim, Parfit notes, we can rule out many potential Selectors through straightforward observation of the world around us. For example, the world obviously does not have the feature of containing the fewest number of things. Contrary to many axiological and theistic theorists, Parfit also considers it evident that the actual world does not have the feature of being the best possible world given how many bad things exist.

But how are we to assess a proposed Selector's descriptive claim in cases where the actual world arguably possesses the special feature identified? Parfit suggests the following guidepost: the easier that it would be to explain the world's having a particular feature, the "more reason to believe that this possibility obtains."[11] He illustrates with the possibility that we live in a multiverse. We should consider the cosmos being a multiverse a viable Selector because that would make it easier to explain the world's compatibility with life. As discussed earlier, contemporary physicists have elaborated the fine-tuning problem, which poses the question of why the world contains just the right conditions for life to emerge even though it was extraordinarily unlikely that this would be the case.[12] If there is only one universe, it is extremely difficult to explain the world's suitability for life. However, some observers have posited the existence of a multiverse, in which countless separate universes have widely varying conditions. In a multiverse, it would be far more likely that at least one universe would have the right mix of conditions to support life. Adopting this reasoning, Parfit contends that a multiverse makes the existence of life easier to explain. As a result, we should favor the descriptive claim that the actual world is a multiverse over the claim that our world is the only one.[13]

Parfit also offers a frame for assessing a proposed Selector's explanatory claim. Assuming that the world has some special feature, the question is

10 Parfit, "Why Anything? Why This? Part 1," 27. Parfit also recognizes the possibility that the world has more than one special feature, each of which contributes to the reason for the world's existence. In that case, the reason for the world's existence would lie in multiple "partial Selectors." Parfit, "Why Anything? Why This? Part 2," 22.

11 Parfit, "Why Anything? Why This? Part 1," 25.

12 See Chapter Five.

13 Parfit, "Why Anything? Why This? Part 1," 25.

whether it just happens to have that feature, or whether that feature is the reason why the world exists in the form that it does. While allowing that the analysis does not lend itself to bright-line rules, Parfit offers a standard for assessing the plausibility of a proposed Selector's explanatory claim. Specifically, Parfit writes, "If we suppose that reality has some special feature, we can ask which of two beliefs would be more credible: that reality merely happens to have this feature, or that reality is the way it is because this way has this feature." If it is more plausible that the feature actually explains reality taking the shape it does, then the feature can be recognized as a "credible Selector."[14]

To illustrate, Parfit gives examples of what he considers credible Selectors. If every possible world existed, it would be more believable that this feature explained the world's existence than that the world just happened to have this feature. The same would be true if the world was the "best, or the simplest, or the least arbitrary," made "reality as full and varied as it could be," or had the most elegant fundamental laws. Each of these features, then, counts as a credible Selector." On the other hand, Parfit asks us to suppose that reality included 58 distinct worlds. With respect to that feature, Parfit says, "it would be more reasonable to believe that the number [of worlds] that existed merely happened to be 58." The feature that reality encompasses 58 worlds is not a credible Selector.[15]

Employing these principles, Parfit makes the case for a particular Selector: the "All Worlds Hypothesis." To convey its meaning, Parfit draws a distinction between "cosmic possibilities" and "local possibilities." A "cosmic possibility" describes "the way that reality might be as a whole." It refers to everything that exists. A "local possibility" describes a way that "some part of reality, or local world, might be." The All Worlds Hypothesis proposes the cosmic possibility that "every possible local world exists."[16] If the hypothesis is correct, it means that the actual cosmos comprises "every conceivable kind of world," including worlds radically different from our own. Indeed, most "other worlds would have very different elements and laws."[17] Along with the descriptive claim that the cosmos includes every possible local world, the All Worlds Hypothesis also makes the explanatory claim that this feature is the reason why the cosmos exists in the form that it does.

Parfit offers several arguments to support the All Worlds Hypothesis's descriptive claim. First, assuming that something like the Big Bang Theory is correct, Parfit sees little reason to think that there has been only one such

14 Parfit, "Why Anything? Why This? Part 2," 22.
15 Parfit, "Why Anything? Why This? Part 2," 22.
16 Parfit, "Why Anything? Why This? Part 1," 25.
17 Parfit, "Why Anything? Why This? Part 1," 26.

origin event. "Most kinds of thing, or event," he notes, "have many instances."[18] As noted, Parfit also considers the fine-tuning problem a reason to favor the view that our cosmic home is merely one universe within a multiverse containing widely varying worlds. Given his aversion to leaving things unexplained, he cannot accept the notion that the universe's fitness for life is nothing but a huge coincidence. Based on the objection from evil, he also rules out axiological and theistic explanations for the world's friendliness to the emergence of life. The All Worlds Hypothesis, however, provides a compelling response to the fine-tuning problem. If every possible world exists, then it is hardly surprising that at least one of them has conditions suitable for life.[19]

Parfit offers another reason to embrace the All Worlds Hypothesis: it dramatically reduces the arbitrariness in the cosmos and, thus, leaves less unexplained than any other cosmic possibility. Consider all of the things that seem arbitrary if our universe is the only one. For instance, why does light travel through a vacuum at a maximum rate of nearly 300,000 kilometers per second? Why isn't the speed of light a little faster or slower than that? And why does light travel faster in a vacuum than through water rather than the other way around? For that matter, why does water have all of the particular properties that it does? And why does the universe have the specific amount of copper that it does, or the specific amount of overall mass and energy? The questions proliferate. If our world is the only one—if it truly is a *uni*-verse—then we are left with an enormous number of unanswered questions.[20] However, if we live in a multiverse, that would remove a great deal of the arbitrariness. Questions taking the form "Why is the world like X rather than Y?" would be based on a false premise. It would not be true that the speed of light was nearly 300,000 kilometers per second rather than 301,000 kilometers per second. Both of those claims would be true, but in different worlds. If each possible feature of reality is actualized in a distinct world, then we no longer need explanations for why our world takes the shape it does. We happen to find ourselves in a world with a particular set of features, and other worlds have different sets of features.[21]

Assuming that the cosmos includes every possible world, should we conclude that it is the Selector? Recall Parfit's touchstone for evaluating a Selector's explanatory claim. We ask whether it is more credible that the

18 Parfit, "Why Anything? Why This? Part 1," 25.

19 Parfit, "Why Anything? Why This? Part 1," 24–25. The use of a multiverse theory to address the fine-tuning problem is discussed in Chapter Five.

20 Granted, physicists might answer questions of this type in terms of the need for consistency between various aspects of contemporary science, but Parfit's point is that there are countless other possible worlds with very different—but self-consistent—scientific regularities.

21 Parfit, "Why Anything? Why This? Part 1," 26.

cosmos accidentally possesses the proposed special feature or, rather, that the feature is the reason for the cosmos existing as it does. Although Parfit does not provide detailed standards for making judgments of this kind, he takes it for granted that some cosmic possibilities are "intrinsically more likely" than others in the sense that they "had a greater chance of being how reality is."[22] He notes that making this kind of judgment is common in science: "When we choose between scientific theories, our judgments of their probability cannot rest only on predictions based on established facts and laws. We need such judgments [about the intrinsic likelihood of particular things being the case] in trying to decide what these facts and laws are." To illustrate, he asks us to compare two cosmic possibilities. One is a "lifeless Universe consisting only of some spherical iron stars, whose relative motion is as it would be in our world." In the other, "things are the same, except that the stars move together in the patterns of a minuet, and they are shaped like either Queen Victoria or Cary Grant."[23] To Parfit, it is evident that it was intrinsically less likely that the cosmos would take the latter form.

Returning to an example noted above, he asks us to assume that the cosmos includes 58 worlds. The number 58 might have somewhat interesting properties, Parfit observes, such as "being the smallest number that is the sum of seven different primes." Nevertheless, the number is not sufficiently distinctive, or, well, special, to think it more credible that the feature of containing 58 worlds was the Selector. By contrast, if reality had included no worlds—if the cosmic reality were a state of absolute nothingness—that would be a feature special enough to make it a credible Selector. The same is true of the All Worlds Hypothesis. In short, all and nothing are special, but the same cannot be said for all of the figures in between.[24]

Reflecting his preference for theories with the broadest explanatory scope, Parfit explores whether any theory could leave nothing unexplained. Suppose that we successfully identify the Selector. A natural question is to ask whether the Selector itself has an explanation.Parfit adopts the term "Higher Selector" for a prior Selector that explains why the Selector is the Selector. Since the Selector is a law that explains why the cosmos takes the form that it does, a Higher Selector is a law that explains why the Selector takes the form that it does. For instance, if the All Worlds Hypothesis is the Selector, perhaps it could be explained by a Higher Selector which requires that the Selector minimize arbitrariness. Thus, this line of inquiry potentially initiates an infinite regress. Recognizing this possibility, Parfit

22 Parfit, "Why Anything? Why This? Part 2," 22.
23 Parfit, "Why Anything? Why This? Part 2," 22.
24 Parfit, "Why Anything? Why This? Part 2," 22.

inquires whether it is possible for there to be a Highest Selector that has an explanation, thus leaving no stray questions unanswered.[25]

Parfit argues that there cannot be such an explained Highest Selector. He considers a number of ways that one might think that the Highest Selector could be explained, and rejects all of them. First, perhaps the Highest Selector could be explained by something outside itself. Parfit rules this out because, by definition, the Highest Selector is not explained by a Higher Selector. The Highest Selector cannot be explained by a Higher Selector, because then *that* would be the Highest Selector. Second, one might suppose that the Highest Selector could explain itself. Parfit rules out this possibility because he rejects the idea of self-explanation. Only something that already exists can furnish an explanation, he argues, and this means that we cannot rely on it as an explanation without assuming its existence from the start, which would be circular.[26] Third, one might suppose that the Highest Selector is grounded in logical necessity. However, Parfit finds logical necessity inapt as an explanation for the Highest Selector because it only excludes possible realities that entail outright inconsistencies. Since this means that logical necessity leaves countless possibilities still in play, it cannot explain why the Highest Selector would take one particular form. Lastly, Parfit denies that causal necessity could explain the Highest Selector, because that kind of necessity depends on the pre-existence of specific physical realities, and, thus, assumes what it is supposed to explain. Unable to identify any viable source of explanation for the Highest Selector, Parfit concludes that if there is a Highest Selector it cannot have an explanation.[27] The Highest Selector would have to be a brute fact. In the end, this is the only necessity that Parfit thinks we can ascribe to the Highest Selector: the conclusion that it must be a brute fact.

Notwithstanding Parfit's preference for theories that explain as much as possible, he acknowledges the possibility that there is no Selector at all, a state of affairs he calls the "Brute Fact View." Indeed, he concedes that there is no way to conclusively disprove the Brute Fact View. Moreover, it might even be difficult for us to tell the difference between living in the reality described by the Brute Fact View and living in the reality described by the All Worlds Hypothesis. In either case, we would likely find the world to be relatively generic rather than exhibiting extraordinary or highly unexpected features. Since the Brute Fact View posits no explanation for existence, there is no reason to think that a world based in randomness would have special

25 Parfit, "Why Anything? Why This? Part 2," 23–24.

26 We saw that Swinburne employs similar reasoning in arguing that the explanation for the world's existence cannot be self-explanatory. See Chapter Six.

27 Parfit, "Why Anything? Why This? Part 2," 23.

features or display remarkable characteristics. Our world would also likely be unremarkable if the All Worlds Hypothesis is correct. The multiverse would include some worlds with astounding features, but these would be vastly outnumbered by more pedestrian ones. After all, the very notion of a world with special or extremely surprising features entails that worlds of this kind are relatively uncommon. Thus, it would be far more likely in a multiverse that we would find ourselves in a relatively generic world.[28]

Nevertheless, even without being able to categorically rule out the Brute Fact View, Parfit thinks that we have better reasons to accept the All Worlds Hypothesis. Two considerations already noted militate in its favor: (1) it provides the best response to the fine-tuning problem, and (2) it eliminates many questions associated with the seeming arbitrariness of our universe's multifarious features. In short, the All Worlds Hypothesis explains much more than the Brute Fact View.[29]

Van Inwagen's Probabilistic Analysis

Peter van Inwagen's approach to the puzzle of existence shares important elements with Parfit's. Like Parfit, van Inwagen aims to provide a reason why there is a physical world through reasoning based in explanatory principles that do not depend on the existence of physical things. Thus, he assumes that we can reason meaningfully about various possibilities for existence even in an imagined context that does not yet contain any existing things. Moreover, both thinkers eschew more familiar approaches to explaining the world's existence.

In addition, while both Parfit and van Inwagen offer their own favored theories, neither maintains that it is possible to definitively establish a particular solution to the puzzle of existence. As we have seen, Parfit contends that ultimate explanation (the Highest Selector) cannot be based in necessity, and his general approach employs various criteria to assess the relative plausibility of proposed theories. In a similar vein, van Inwagen maintains that the reason for the world's existence cannot be true by necessity. His argument for this position draws on the distinction between concrete objects (the physical things we observe in the world) and abstract objects

28 Parfit notes that there is an exception. Our universe's suitability for life may be a special feature. However, Parfit does not consider this relevant to deciding between the Brute Fact View and the All Worlds Hypothesis because "it is one that we cannot help observing." Given that we are here to discuss the question, it is necessarily the case that we live in a world with that feature. Parfit reasons that the only special features that could be considered evidence against the Brute Fact View would be those that are not necessarily linked with our universe's fitness for life, and he does not think that our universe evinces such special features. Parfit, "Why Anything? Why This? Part 2," 22–23.

29 Parfit, "Why Anything? Why This? Part 2," 24.

(such as numbers, which do not occupy a particular place in space and time). Crucially, van Inwagen accepts a Kantian claim that we examined earlier: that necessity is not a property that can be attributed to the existence of concrete things.[30] This means that the idea of necessary existence is only valid with respect to abstract objects. Since the physical world comprises concrete objects, van Inwagen argues that there cannot be a necessitarian explanation for its existence.[31]

Notwithstanding the parallels, van Inwagen offers an account that differs significantly from Parfit's. We saw that Parfit extracts a general explanatory structure from a more familiar approach to explaining the world's existence (axiological). Van Inwagen instead begins with a set of premises that serve as the basis of his argument. He seeks to demonstrate not that the world had to exist, but, rather, that nothingness is "as improbable as anything can be."[32] If he can establish that the non-existence of physical things was almost impossible, this amounts to showing the physical world's existence was extraordinarily likely. Before examining the argument in detail, we can summarize van Inwagen's account in a sentence: The physical world exists because there are so many more possible worlds that contain physical things than those that do not. (More specifically, there are infinitely many possible worlds that include physical existence, and only one possible world that contains no physical things at all.)

Van Inwagen's argument rests on four premises. Least controversial is the first premise, which observes simply that "there are some beings." In other words, the actual state of affairs is not one of absolute nothingness. The second premise holds that "if there is more than one possible world, there are infinitely many."[33] To support the claim that there is more than one possible world, van Inwagen notes that if there is only one possible world, then the only possible world is *our* world, which is tantamount to saying that it exists necessarily. However, for reasons outlined above, van Inwagen rejects the notion that the world exists necessarily. But why does van Inwagen assume that if there is more than one possible world, then there are infinitely many? Once we allow for any differences at all between possible worlds, he contends, it would be "bizarre" to suppose that there could be a limited number of permissible variations in the properties that things have. The properties that we observe in the things around us "seem to imply a myriad of dimensions along which [they] could vary continuously."

30 See Chapter Six.

31 Peter van Inwagen, "Why Is There Anything at All?," *Proceedings of the Aristotelian Society* 70 (1996): 95–96.

32 Van Inwagen, "Why Is There Anything at All?," 99.

33 Van Inwagen, "Why Is There Anything at All?," 99.

Consequently, we have no reason to think "that there might be just two or just 17 or just 510 [possible] worlds."[34]

The third premise states that "there is at most one possible world in which there are no [concrete] beings"[35] This premise is rooted in van Inwagen's view that, unlike concrete objects, abstract objects (such as numbers and other mathematical objects) exist necessarily. To acknowledge the necessity of abstract objects is to say that they do not vary from one possible world to the next. They are the same in all possible worlds. Suppose, then, that two worlds both have no concrete beings. We have just said that they cannot have different abstract objects, and they obviously cannot have different concrete beings (since they do not have any concrete beings at all). This means that there is no way in which two worlds with no concrete beings could differ. The upshot is that there is only one possible world that contains no physical things.[36]

The fourth premise maintains that "for any two possible worlds, the probability of their being actual is equal."[37] As van Inwagen acknowledges, this "equiprobability thesis" is likely the most controversial of the four premises. Notably, the equiprobability thesis is a direct repudiation of Leibniz's assumption that absolute nothingness would have been the simplest and easiest form for reality to take. Leibniz urges us to feel awe at the world's existence because it is surprising that there is a world at all.[38] For Leibniz, the simplest and easiest possible reality is one in which nothing exists. In effect, Leibniz's position is that the absence of anything physical was more likely than the existence of a physical world, which prompted him to inquire why there was something rather than nothing. Van Inwagen rejects this view. Leibniz's claim is groundless, van Inwagen argues, because in a pre-existential context there is no frame of reference for assessing one possible state of affairs as more or less likely than any other. To be sure, we could make sense of the idea that nothingness is a favored condition by presuming the existence of something or someone with various tastes or preferences, such as a deity or a universe-selection machine. Van Inwagen resists such a move, however, contending that it would be circular to assume the existence of an entity or being that makes choices about whether anything should exist. Thus, in an imagined context in which nothing yet exists, all

34 Van Inwagen, "Why Is There Anything at All?," 100–01.
35 Van Inwagen, "Why Is There Anything at All?," 99.
36 Van Inwagen, "Why Is There Anything at All?," 101.
37 Van Inwagen, "Why Is There Anything at All?," 99.
38 See Chapter Six.

possible worlds have an exactly equal chance of being realized, including a state of absolute nothingness.[39]

Together, then, the four premises indicate that there were an infinite number of equally probable worlds, and only one of them lacked physical existence. This effectively amounts to saying, van Inwagen concludes, that the probability of there existing no physical beings is zero. Van Inwagen's answer to the question of why there is something rather than nothing, then, is that the odds are stacked in its favor by an extraordinary margin.[40]

Smith's "Law of the Simplest Beginning"

Quentin Smith is particularly explicit in calling his theory "nomological" and titling the ultimate reason why there is a physical reason as a law. Like Parfit and van Inwagen, he grounds the world's existence in explanatory laws that do not depend on anything physical for their reality. However, he arrives at his account through a distinctive path. Where Parfit excavates a general explanatory structure from his analysis of competing approaches, and van Inwagen derives a probabilistic conclusion from a series of premises, Smith's jumping off point is the Big Bang Theory. Smith seeks an ultimate explanation that is informed by the best science, but he also recognizes that science alone is inadequate because it would be circular to identify something physical as the reason for the world's existence. Thus, he aims to identify something that is not itself a physical thing that can explain the Big Bang. He finds that bedrock explanation in the "Law of the Simplest Beginning" (or LSB), which holds that "the simplest possible thing comes into existence in the simplest possible way."[41] Working from this founda-

39 Van Inwagen, "Why Is There Anything at All?," 107–08. While van Inwagen maintains that each possible world had an equal probability of being actual, he does not offer an account of how or why one of the possible worlds in particular is actual.

40 Van Inwagen, "Why Is There Anything at All?," 113–14. There is an aspect of van Inwagen's argument that may seem perplexing. He maintains that the infinity of possible worlds and the equiprobability thesis imply that each possible world had a zero probability of being actual. However, this means that the existing world that is our home, too, had a zero probability of being actual, which may seem to entail a blatant contradiction. See E.J. Lowe, "Why Is There Anything at All?," *Proceedings of the Aristotelian Society* 70 (1996): 113–14. Van Inwagen addresses this potential confusion by noting that in the context of his favored understanding of probability theory, while the "probability of any impossible event is 0, … not all events whose probability is 0 are impossible." Van Inwagen, "Why Is There Anything at All?," 99 n. 5. Accordingly, van Inwagen does not assert that an absence of physical existence was impossible, but, rather, that it was "as improbable as anything can be." Van Inwagen, "Why Is There Anything at All?," 99. Putting aside technical or controversial aspects of probability theory, we can understand the intuitive gist of van Inwagen's argument as holding that the chance of any single possible world being actual was extremely unlikely, so if there was only one possible world with no physical things and countless worlds with physical things, then it is hardly surprising that the actual world includes physical things.

41 Quentin Smith, "Simplicity and Why the Universe Exists," *Philosophy* 72 (1997): 128.

tional law, Smith identifies the "simplest possible thing" as the singularity that is the Big Bang Theory's origin point.

In rooting existence in the law that the simplest thing comes into being in the simplest way, Smith's theory naturally places great importance on the meaning of simplicity. For Smith, the simplicity of a *thing* is judged by its size, mass, and duration in time. Specifically, the simplest possible thing would take up no space, have no mass, and manifest zero duration in time. In addition, it would instantiate the fewest laws and have the fewest positive properties. The simplest *way* for a thing to originate is for it to "have no positive relations to any grounds for coming into existence, ... to exist from nothing (from no previously existent material), ... to come to exist by nothing ... and to come to exist for nothing (for no purpose)."[42] In sum, Smith's understanding of simplicity minimizes identifiable properties both with respect to the nature of a thing and the process by which it came to be.[43]

How can Smith connect ultimate explanation with science through a master law that shuns positive properties? He endorses an interpretation of the Big Bang Theory—embraced by a good number of physicists—that the Big Bang originated in a singularity. In this view, the singularity that generated the Big Bang was "spatially and temporally pointlike," meaning that it had no spatial dimensions and existed for only an instant before producing the explosion that set the universe's expansion in motion.[44] The singularity did not itself contain mass, but it gave rise to a universe that did. In Smith's telling, the singularity was governed by no laws—other than the LSB—and it exhibited no properties other than that of being the simplest possible thing. It is the singularity's dearth of properties and governing principles that allows Smith to say that the universe came into being from the simplest possible thing in the simplest possible way.

According to Smith, the singularity's existence is explained by the LSB, but the singularity itself is lawless, meaning that it could lead to any outcome with equal probability.[45] Out of the countless possibilities, the singularity randomly generated our universe's actual features. The randomness of the process is critical to Smith's theory. If it was not random, then the singularity

42 Smith, "Simplicity and Why the Universe Exists," 129.

43 The contrast with Parfit in terms of the foundational assumptions is striking. Where Parfit's preference for leaving the least unexplained leads him to posit an ontologically extravagant cosmos encompassing every possible world, Smith's distinctive conception of simplicity leads him to view the world as sourced by an origination as close to nothingness as possible.

44 Smith, "Simplicity and Why the Universe Exists," 129.

45 Smith, "Simplicity and Why the Universe Exists," 129.

would need definite properties to pick out a particular world from all of the possibilities, and that would mean that it fails to satisfy the LSB.[46]

We have repeatedly encountered the fine-tuning problem: the mystery posed by contemporary physics as to why the universe has just the right mix of features to support life. We have also noted a variety of responses to the problem, including the view that the cosmos is inherently directed toward goodness, and various multiverse theories. While Smith takes the fine-tuning problem seriously, he endorses a distinctive response: that the universe's fine-tuning for life is the result of a huge coincidence. The response may seem surprising in light of Smith's affinity for cutting-edge scientific theories and science's reluctance to recognize coincidence as the explanation for unexpected observations. However, Smith is driven to this position by his reasons for rejecting the alternatives. He cannot accept answers that view the universe's fine-tuning as resulting from a cosmic directedness toward the good because the LSB has no place for assessments of value. Neither can he accept multiverse theories because they flout the maximal simplicity required by the LSB. Seeing no other options, Smith bites the bullet, maintaining that our existence is the result of a very lucky break.[47]

In making the case for his theory, Smith contends that the LSB and the Big Bang Theory work hand in glove. The LSB predicts the Big Bang singularity, because the latter fits Smith's understanding of simplicity to a tee. At the same time, scientific evidence supports the Big Bang singularity, which serves as confirmation of the theory's predictions. Moreover, the case for the LSB is strengthened by the observation that, without it, we would not expect a Big Bang singularity. As Smith puts it, the LSB explains the singularity, "for explanation is the converse of confirmation; if something confirms a law, then the law explains that thing."[48]

The randomness that characterizes the singularity in Smith's theory is distinct from the randomness that contemporary science recognizes in the occurrence of events at the quantum scale. As discussed earlier, a prevailing interpretation of quantum theory (known as the "Copenhagen interpretation") holds that certain events occur in a manner that is genuinely random.[49] As understood in the Copenhagen interpretation, physicists can model events (such as an atom's emission of radiation) according to statistical probabilities with great precision, but it is impossible to predict the precise moment when a particular atom will emit radiation. The important

46 Although analogous in certain ways, the randomness behind the singularity's generation of a particular universe from all the possibilities is not the same as the randomness that characterizes the operation of our existing universe. See Chapter Three.

47 Smith, "Simplicity and Why the Universe Exists," 125–26, 129.

48 Smith, "Simplicity and Why the Universe Exists," 129.

49 See Chapter Three.

point is that quantum theory describes the way that physical things operate in our universe. When Smith talks about the random manner in which the singularity led to the universe's existence, he is not referring to quantum theory. The singularity was not constrained by the patterns described by quantum theory. Rather, the singularity existed *prior to* the universe's existence. Our universe might not have ended up with the patterns of behavior described by quantum theory. It might have had very different physical things behaving according to different patterns of behavior. In Smith's theory, the singularity generated our universe—with all of its particular features—in an unconstrained and random manner. This is pivotal to Smith's claim that the singularity satisfies the LSB's requirements, because the singularity's lack of positive properties makes it extraordinarily simple. It is the explanatory primacy of a non-physical law governing existence that makes the nature of the ultimate explanation in Smith's account nomological rather than scientific. In Smith's words, the LSB is "a nomological explanation," in that "the singularity is subsumed under a law that provides an answer to the question 'why does the singularity exist?'"[50]

To bolster his theory, Smith aims to show that it offers a simpler explanation for the world's existence than theism. For Smith, coming into existence in the simplest possible way means coming into existence without a cause. Theistic narratives, on the other hand, posit a God who caused the world to come into existence through intentional actions. As a result, theistic accounts of the universe's origins cannot satisfy the LSB. Earlier, we discussed Swinburne's case for theism based in his definition of God as the simplest possible person, and his argument that theism offers the simplest explanation for the world's existence.[51] Smith and Swinburne both place central importance on simplicity yet they reach contrary conclusions owing to their clashing definitions of simplicity. Since Swinburne assesses simplicity according to the number of explanatory factors required, theism's need to posit only one being makes it the simplest possible explanation. By contrast, Smith's LSB calls for an origin point that lacks properties, laws, and causes, which is best fulfilled by the Big Bang singularity.

Nevertheless, there is an interesting point on which Smith and Swinburne (as well as Parfit and van Inwagen) converge: the claim that ultimate explanation is not necessary in character. Just as Swinburne contends that God's existence has no explanation, Smith contends that the LSB must be recognized as a brute fact rather than as something for which a reason can be identified.

50 Smith, "Simplicity and Why the Universe Exists," 129.

51 See Chapter Six.

Nozick's "Assumption of Fecundity"

In his 1981 book *Philosophical Explanations*, Robert Nozick devotes a chapter to asking how close an explanation for the world's existence can come to leaving nothing unexplained. What makes Nozick's discussion particularly salient is that he approaches the question of why the world exists as, above all, a quest to identify explanatory laws. While Nozick's imaginatively sprawling discussion touches on a wide range of topics, it is not principally directed toward science, mathematics, axiology, or theism as frames for explaining the world's existence. Rather, the central motivating questions concern general principles explaining what shall exist.

In probing the possibility of an ultimate explanation that leaves nothing unexplained, Nozick considers whether an ultimate explanation could be self-subsuming. A law (call it "Law S") is self-subsuming if it holds that all laws with a certain property are true, and Law S itself has that that same property. Consider a statement like the following: "All claims with seven words are true." The law governs what qualifies as true while providing a reason for its own truth. Nozick concludes that not even a self-subsuming law can answer every question. In particular, it leaves open the question of why reality takes a form that can be explained by a self-subsuming law.[52]

While denying the possibility of complete explanation, Nozick nevertheless proposes a law ("the assumption of fecundity") that could explain all things that can be explained. We have seen that Leibniz motivates the question of why there is something rather than nothing with the claim that nothingness would have been the easiest and simplest possible state of affairs. While many have embraced Leibniz's framing of the question (including Parfit, for example), others have dissented. Van Inwagen bases his probabilistic analysis on the "equiprobability assumption," which holds that every possible world had the exact same chance of being realized. Nozick, too, rejects Leibniz's claim. Indeed, he bemoans the pervasive impact that Leibniz's inegalitarian assumption has had on discourse addressing the reason for the world's existence.[53] One reflection of that impact is the common adoption of Leibniz's phrasing: "Why is there something rather than nothing?" Nozick observes that to ask why some condition (call it "X") prevailed rather than an alternative condition (Y) is to imply that Y was more to be expected. The very wording of Leibniz's inquiry smuggles in the assumption that nothingness was a preferred state. Nozick uses the term "inegalitarian" for views, like Leibniz's, which hold that some possible worlds were more likely to be realized than others. Finding no basis for inegalitarianism,

52 Robert Nozick, *Philosophical Explanations* (Harvard University Press, 1981), 120.

53 Nozick, *Philosophical Explanations*, 121.

Nozick frames his quest for a maximally explanatory answer to the puzzle of existence around the following question: What are the implications of taking seriously the egalitarian proposition that all possible worlds were equally likely?[54]

To explore that question, Nozick suggests a novel way to achieve true egalitarianism in our thinking about the puzzle of existence. We should presume that every possible world—including even the possibility of absolute nothingness—is actual. There are similarities between Nozick's assumption of fecundity and Parfit's All Worlds Hypothesis, which is not surprising when we recall Parfit's preference for theories that leave as little unexplained as possible. Both thinkers hypothesize a cosmos that includes all possible worlds and contend that recognizing such a multiverse minimizes the aspects of reality that remained unexplained. Faced with questions in the form, "Why is reality like X rather than Y," they take the same approach: denying the question's premise. Reality does not have the feature X rather than Y. Instead, it has both features, with each one manifested in a distinct universe. We just happen to be in a corner of reality that has the feature, giving the misleading impression that this feature is invariable throughout the cosmos. Thus, both Nozick and Parfit posit countless, separate, non-interacting worlds in order to remove as much as possible from the scope of the unexplained.[55]

However, there is also an important difference: Nozick's multiverse includes an empty world, while Parfit's does not. The difference results from the two thinkers' disagreement over whether nothingness is a more expected state of affairs. Parfit accepts the inegalitarian position that Nozick so strenuously rejects. In Nozick's view, to exclude the null world (containing nothing) from the multiverse is to illegitimately make room for the inegalitarian question of why there is something rather than nothing. To shake off inegalitarianism once and for all, Nozick envisions that reality encompasses *every possibility*. This means that the cosmos includes not only all of the possible worlds that contain physical things but also a world in which nothingness prevails. Nozick wants to suppress *all* questions that take the inegalitarian form, "Why X rather than Y?" and we can only do that by recognizing the existence of a world of absolute nothingness. Thus, if someone asks why there is something rather than nothing, the proper response is: "There's not. There's both."[56] Nothing and something are both actual, but in separate, non-interacting worlds. Driven by the aim of developing a genuinely egalitarian approach to the puzzle of existence, Nozick proposes a cosmos so inclusive that it simultaneously encompasses everything and nothing.

54 Nozick, *Philosophical Explanations*, 128.
55 Nozick, *Philosophical Explanations*, 128.
56 Nozick, *Philosophical Explanations*, 130.

Challenges for a Nomological Approach

The nomological theories that we have examined differ in many ways. As a result, each is subject to objections that are peculiar to the individual theory. We will briefly consider examples of such objections before examining general challenges for nomological theories based in their shared features.

To begin, we consider a potential objection to the reasoning that Parfit uses to establish his All Worlds Hypothesis. Having contended that we have good reason to suppose that we live in a cosmos encompassing every local possibility, Parfit aims to show that this feature is the reason for reality taking the form that it does. In doing so, he outlines a standard for assessing explanatory claims generally. According to this standard, we should be more willing to accept an explanatory claim if it seems unlikely to have come about by chance. To show that this standard militates in favor of the All Worlds Hypothesis, he contrasts it with another global possibility: namely, one encompassing exactly 58 different local possibilities. Assuming that we live in such a reality, Parfit argues that the number 58 is not special enough to count as a credible Selector. By contrast, for Parfit, the property of comprising all possible worlds is much more special. Thus, if we live in such a reality, it would be reasonable to recognize that feature as explaining the world's existence.

It is arguably a difficulty for this line of argument that it lacks a sufficient basis for making determinations about which global possibilities count as more or less special than others. Parfit seems to take it for granted that we can make comparative assessments of how surprising global possibilities are. But it is far from clear what grounds we might have for making such assessments. Parfit does not claim that less credible Selectors entail contradictions or otherwise violate the requirements of logic, nor does he identify mathematics or anything else as providing a scale of amazingness versus ordinariness. Moreover, one might object that notions like surprise and astonishment only take on content in the context of our experience within an existing world in which we find ourselves more likely to encounter some kinds of things than others. If that is so, then it may undermine Parfit's attempt to draw on such concepts in making the case for why the world has the particular properties that it does in the first place.

Turning now to Smith's nomological theory, recall that he finds ultimate explanation in the Law of the Simplest Beginning. This law is independent of the physical world and is not itself an expression of scientific theories. At the same time, Smith's theory draws on contemporary science in arguing that the the singularity—which satisfies the LSB—engendered the universe and set in motion events described by the Big Bang Theory.

The theory incorporates scientific claims, including some controversial ones on which contemporary researchers disagree. Smith's reliance on contested scientific propositions subjects his theory to potential objections from those with opposing views. For instance, the philosopher of science Robert Deltete challenges the scientific assumptions underlying Smith's arguments. Using quantitative models, Big Bang theorists effectively wind the clock back to estimate the universe's conditions at the earliest stages.[57] However, the relevant equations break down as the models are used to calculate the universe's conditions at times closer and closer to an imagined earliest event. This makes it impossible to definitively predict what would arise from such an event. It is common ground that the equations break down, but scientists disagree over how to interpret this. Smith adopts a particular interpretation holding that the singularity represents a starting point from which virtually anything could emerge. Deltete notes, however, that Hawking and other prominent physicists advance an opposing interpretation: that general relativity cannot be a complete scientific framework for understanding cosmic origins.[58] Interestingly, then, while the most fundamental explanation in Smith's theory is not a scientific law, the success of his account may nevertheless turn to some degree on contested scientific questions.

A different objection to Smith's theory concerns his description of the singularity, which is the origin point for the Big Bang. In Smith's theory, the world's existence is explained by the LSB, which maintains that the simplest possible thing comes into existence in the simplest way. The simplest possible thing has the fewest properties, has the least mass, size, and duration of existence, and instantiates the fewest laws. The simplest possible way for something to come into existence is to have no cause or purpose for existing.[59] In contending that the singularity satisfies the LSB, he characterizes the singularity as having zero mass and size, and existing only "for an instant before exploding with a 'big bang.'"[60] Because the singularity is lawless, causeless, and purposeless, it could lead to any kind of universe. The process through which the singularity generates the universe is random. Our universe could have been very different, and there is no reason why it is this way rather than another. Smith's description of the singularity is almost entirely negative. It tells us what the singularity is not and what it lacks. There is one big exception: the singularity exploded with a big bang, thus giving rise to the universe. In sum, the singularity lacks positively describable characteristics other than the ability to produce a world.

---

57 The Big Bang Theory is discussed in Chapter Three.

58 Robert Deltete, "Simplicity and Why the Universe Exists: A Reply to Quentin Smith," *Philosophy* 73, no. 285 (1998): 492–94.

59 Smith, "Simplicity and Why the Universe Exists," 129.

60 Smith, "Simplicity and Why the Universe Exists," 125.

A possible difficulty is that the thorough manner in which Smith's theory deprives the singularity of positively describable features undermines its intelligibility. We may wonder about the extent to which the singularity effectively explains the world's existence if the only thing that we can say about it is that it explains the world's existence. Without more, the singularity may seem to work more as a placeholder than as an identifiable explanation. Smith is at pains to show that his theory provides a more satisfying explanation for the world's existence than theism. Interestingly, theistic accounts of the universe's origin are sometimes criticized on comparable grounds. As we saw in the last chapter, critics have charged theism with incoherence, alleging that it simply substitutes one notion shrouded in mystery (the idea of God) for another (the reason for the universe's existence). Though for different reasons, Smith's theory is potentially vulnerable to a similar critique.

Another objection to a particular nomological account arises from a critical way in which Nozick's "assumption of fecundity" differs from Parfit's All Worlds Hypothesis. The distinction is particularly intriguing in light of how much the two thinkers share. Both Nozick and Parfit place great importance on leaving as little unexplained as possible. Parfit's criteria for assessing proposed theories favor those that leave the least unexplained, and Nozick frames his inquiry as an exploration into how close we can come to explaining everything. Moreover, both thinkers are led by their emphasis on explaining as much as possible to endorse theories positing a multiverse encompassing every possible world. The connection between a multiverse and maximizing explanation concerns the reduction of cosmic arbitrariness. If there is only one universe, then we are left with countless unanswered questions about why the features of the universe are one way rather than another. However, in the infinitely varied multiverse it is not true that things are one way rather than another. Every possible way for the world to be is a reality in at least one member of the multiverse. Thus, the number of unanswered questions about apparently arbitrary things in the world is drastically reduced.

Notwithstanding their similarities, Nozick and Parfit diverge on a critical issue. Like Leibniz, Parfit motivates his quest for an explanation of the world's existence by assuming that nothingness would have been the simplest and easiest state of affairs. He thus pursues the Leibnizian question of why there is something rather than nothing. Nozick, however, repudiates Leibniz's "inegalitarian assumption." He explores the implications of the most rigorous version of egalitarianism possible, which he takes to mean that the cosmos encompasses every possible world, *including an empty world*.[61] Thus, unlike Parfit's multiverse—which includes all possible worlds populated by physical things—Nozick imagines a multiverse that

61 Nozick, *Philosophical Explanations*, 130.

also contains a world with nothing at all. Nozick rejects the question of why there is something rather than nothing, because its phrasing smuggles in the inegalitarian assumption. Only a multiverse that includes the null world finally shakes off the inegalitarian assumption, which Nozick views as unsupported and illegitimate. This move allows Nozick to offer a distinctive response to the question of why there is something rather than nothing: there is *not* something *rather than* nothing. There is both something *and* nothing, although they are instantiated in separate, parallel worlds.

By including an empty world in his multiverse, Nozick opens up his account to a potential objection. In particular, one might question the consistency of asserting that absolute nothingness exists within a cosmos that also includes physical existence. Let us assume for the sake of argument that Nozick's fecundity assumption is otherwise successful in eliminating any trace of inegalitarianism. Suppose, for example, that Nozick's multiverse eliminates questions about why light travels at one speed rather than another. That is, reality includes not only our world—in which the speed of light in a vacuum is nearly 300,000 kilometers per second—but also distinct worlds where the speed of light is faster or slower than that. Likewise, reality includes worlds that vary with respect to the number of atomic elements, or the relative distribution of things with a red, green, yellow, or other color. Differing worlds of this kind do not seem to entail a contradiction. We can imagine them coexisting, and do not know of a reason why that would not be possible. But is it equally coherent to posit absolute nothingness as an actualized reality within a cosmos that also includes physical existence? The other examples concern variables that seem like they could unproblematically be localized in coexisting worlds. By contrast, absolute nothingness by its nature seems to characterize reality as a whole. It is not clear what it would mean to say that there is absolute nothingness over here but not over there. Nothingness does not seem like the sort of thing that could have a particular location. Since Parfit embraces an inegalitarian assumption, he does not need to include the reality of nothingness in his All Worlds Hypothesis, thus avoiding a difficult question raised by Nozick's assumption of fecundity.

Having considered examples of objections aimed at particular accounts, we now examine potential difficulties with nomological theories more broadly. One challenge comes into view when we recall the general nature of a nomological approach and what makes it appealing. Nomological theories are based in explanatory laws that are claimed to be the reason why there is a physical world. These laws are prior to and independent of the world and bear responsibility for its existence. Nomological theorists draw on the idea of possible worlds in the sense of positing that reality could have taken many forms, and foundational laws explain why the world exists as it does. In identifying the explanatory laws that ground the world's existence, nomological accounts reject the alternative approaches that we examined in earlier chapters. Thus, in nomological accounts, the world is not rooted in scientific theories, mathematical objects, goodness, or the actions of a (divine) person. Instead, the explanatory laws are typically treated as sui generis or freestanding, addressing directly the question of what shall exist.

These shared features enable nomological theories to sidestep a range of potential difficulties. The prior and independent nature of the explanatory laws enables nomological theories to avoid circularity. Since the world is fundamentally separate from the laws that explain its existence, we need not worry that these laws illegitimately assume physical existence from the start. Additionally, in establishing a distinctive approach, nomological theories avoid a variety of difficulties facing the other theories that we have examined. For example, challenges confronting a mathematical approach that are tied to its absolute necessitarianism do not arise for nomological theories, because they do not claim to be grounded in logical or mathematical necessity. Nomological theories also do not encounter the problem of evil—which poses such a thorny problem for axiological and theistic approaches—since their explanatory laws are not based in value-laden assessments or teleological considerations. Nor does a nomological approach confront theism's difficulties in reconciling superlative attributes with personhood, since its bedrock principles are impersonal and do not flow from the will of an intentional actor. Nomological theories clear away a thicket of longstanding complications to start afresh. By positing freestanding principles unburdened by baggage associated with other approaches, nomological accounts aim to offer more satisfying responses to the puzzle of existence.

However, as is often the case in philosophical inquiries, it is the appealing features of a nomological approach that introduce challenges at the same time. While scientific, mathematical, axiological, and theistic approaches have their own distinctive vulnerabilities, they also enjoy a significant common advantage over a nomological approach: each one is rooted in a type of explanation with which we are already intimately familiar. After all, our everyday experience is pervaded by the impact on us of scientific laws and quantitative relationships, and we cannot navigate the world without awareness of value considerations and the efficacy of our own intentional actions. The ubiquity of science, mathematics, value, and intentional action in our lives means that we are highly accustomed to the manner in which they provide reasons for actions, events, or states of affairs. Thus, long before we reflect on the reason for the world's existence, we already have reasons to accept science, mathematics, value, and personhood as sources of valid explanations. The core question raised by theories based in those forms of explanation is whether they are apt for providing ultimate explanation.

By contrast, we meet the laws posited by nomological accounts for the first time in the context of explaining the world's existence. While this enables them to shed baggage associated with more familiar approaches, it also raises questions about whether they have the requisite explanatory resources. They do not come to us with the certification of longstanding familiarity. The positing of laws whose unique function is to direct what shall exist may seem ad hoc and, thus, arbitrary. A central challenge for nomological theories is to show why we should accept their asserted principles as capable of being genuinely explanatory in any context at all.

A related question concerns whether the laws purportedly responsible for the world's existence retain a role once they have managed to get everything going. If these laws do retain some sort of potency, then we will want to know in what sense that is the case. If they make no impact once the world is underway, then we may be suspicious of precepts that serve the most significant role imaginable only to then permanently exit the stage. Nomological accounts might begin to overcome these challenges by more fully addressing the relation between the existing world and the laws purportedly responsible for originating it.

Lastly, another objection to nomological theories concerns their plausibility. For some, it may be hard to imagine how principles that are entirely prior to and independent of the world could be responsible for its existence. To be sure, in modern times we have long grown accustomed to the idea of scientific natural laws that characterize the way that objects in the world around us behave. Yet one may incorporate scientific natural laws into the way we think about the world without necessarily assuming that these laws are prior to and independent of anything in the world. Indeed, many think of scientific natural laws as means that we have devised to describe the behavior of things that already exist. That is, one may accept the validity of scientific natural laws without supposing that these laws were themselves responsible for bringing the world into existence. In a different vein, many have no difficulty in accepting that logical laws (like the idea that two contradictory things cannot both be true at the same time) can help us make sense of things in the world even though these laws are not themselves physical objects. However, nomological theories ask us to accept not only that independently true principles help us understand the world, but that these principles explain why there is a world at all.

Of course, a theory's plausibility is something for each person to reflect on and evaluate.[62] Nomological theories have themselves been proposed by thinkers who found the other major theories untenable or implausible.[63] When it comes to the most fundamental questions, we have no guarantee of finding answers that address all of our concerns without also including elements that strike us as weird. Nomological theories contend that they offer the most satisfying and persuasive explanations of the world's existence. The degree of their success is ultimately a matter for everyone's own assessment.

Living in the Lawful Universe

Because nomological theories posit freestanding principles that directly govern what shall exist, they are less suggestive than the other approaches in regard to their implications for the nature of the world and our place within it. Nevertheless, common features of nomological theories allow us to make certain observations. To begin, nomological accounts are built on a core belief that there is a rational explanation for the world's existence. Additionally, they insist that we can discern a good deal about ultimate explanation by reflecting on the requirements of valid explanation. Nowhere is this commitment to rational comprehensibility stronger than in the framework of analysis offered by Parfit, which adopts as a guiding methodological principle that we should prefer answers that leave the least unexplained. A similar spirit is evident in Smith's view that the Big Bang singularity cannot be an ultimate explanation because it is circular to posit a physical entity as the reason for the world's existence. Recognition of our ability to reason about the world's provenance is manifest as well in nomological theories' analysis of possibilities for existence. In different ways, all of the nomological accounts considered rest on the belief that we can talk meaningfully about different sorts of possible worlds even in an imagined context in which nothing yet exists. Responsibility for the world takes the form of laws that mark out certain worlds for realization. Thus, lawlike principles open to our understanding bridge the gap between possibility and actuality, suggesting that reason, broadly understood, has primacy over the fact of physical existence.

At the same time, nomological theories also recognize certain limitations on the extent to which the world's physical existence can be explained. For instance, despite Parfit's aspiration to leave the fewest questions

62 Questions about the weight that we should place on whether a theory sounds "very weird" in considering theories are discussed further in Chapter Nine.

63 See this chapter's opening section.

unanswered, he nevertheless argues that explanation cannot go all the way down. The world's existence can be explained by the Highest Selector, but the Highest Selector itself must be a brute fact. While both van Inwagen and Smith propose accounts explaining why the world exists, they insist that the process responsible for actualizing one world from all of the possibilities is random in character. This means that there can finally be no reason given for why the world takes the particular form that it does. Nozick, too, argues that it is not possible for any account of reality to explain everything; there will always be something left for which we can find no reason.

Nomological theories, then, provide us with a nuanced view of our capacity to answer the most basic questions. We can meaningfully pursue knowledge about the fundamental principles responsible for existence, even though the foundation is not rooted in the kind of necessity that provides self-explanation. While we can aim to show that a particular theory is the most compelling, we cannot demonstrate the reason for the world's existence with deductive certainty.

A nomological approach counsels humility in another crucial respect: it denies that the cosmos revolves around us, our existence, or our interests. The most basic truths in the nomological universe are principles that do not speak in the language of value, personhood, or an inherent directedness toward worthy ends. Nomological principles are responsible for the world, but not in a way that shows us any special solicitude. In this vein, both Parfit and Nozick envision a cosmos in which every possible world exists. Our particular world, then, is not suitable for life as the result of any cosmic purpose. It was just that at least one possible world was bound to have that characteristic, and that happens to be the cosmic possibility that we occupy. In the accounts offered by van Inwagen and Smith, the cosmos's unconcern for us takes a different form: randomness. The world could just as well have turned out in any number of other ways. That it would be hospitable to our existence was a result of mere chance. The nomological universe may provide us with the tools to understand why *it* is here, but it is not so forthcoming with respect to why *we* are.

Questions for reflection and discussion

1. Should we assume that the world is ultimately open to our understanding?
2. Is the world's existence a surprising fact that cries out for explanation?
3. If there is a multiverse, what are the most distinctive features of *our* universe?
4. Is it more likely that the cosmos contains infinite worlds than that it has exactly 19, or 173?
5. What are the strongest arguments supporting Parfit's "All Worlds Hypothesis"?
6. If there is more than one possible world, does this mean that there must be infinitely many?
7. Could the world be the product of entirely random processes?
8. Could there be an explanation for *everything* that would leave no unexplained loose ends?
9. Would there still be arbitrariness in a cosmos that included every possible world?
10. Could an infinitely diverse multiverse include a world with nothing in it at all?
11. Do the ideas of probability or specialness have meaning in an imagined state of nothingness?
12. Is it plausible that the reason for the world's existence takes the form of a law or principle?
13. What implications do nomological theories have for our nature and place in the cosmos?

8. Is the Question Legitimate?

We began by discussing the nature of the question that is our central focus: why is there a physical world? The last five chapters have examined different ways of seeking to answer that question. But there are other manners of responding to the question, such as rejecting the entire inquiry as illegitimate or dismissing it as superfluous. Questions are invitations to pursue answers, but who is to say that everyone must accept the invitation?

The subject of this chapter is a question about questions. When does the pursuit of understanding pass beyond sensible investigation into improper inquiry? Answering this implicates broad issues about the nature of understanding, because the constraints on what count as legitimate questions are wrapped up with the constraints on what count as legitimate answers. After all, one might conclude that an inquiry is illegitimate if it seeks an explanation where the very nature of the question eliminates the possibility of a satisfying explanation. Thus, debate over the legitimacy of asking why the world exists connects with debates over what count as acceptable forms of explanation. In this chapter, we will consider challenges to the puzzle of existence. First, we examine challenges that stress the distinction between individual objects and the totality of all existing things. We then consider two prominent worldviews that render the puzzle of existence—at least in the form we have posed it—inappropriate or beside the point: Nāgārjuna's *Madhyamaka* Buddhism and Śaṅkara's Advaita Vedānta Hinduism.

Objections to Seeking Reasons for the Totality of Existing Things

While scholars have objected to asking why the world exists for a variety of reasons,[1] we begin with a particularly prominent argument associated most famously with the eighteenth-century philosopher David Hume,[2] as well as more recent thinkers like Paul Edwards and Bertrand Russell. The objection concerns the difference between individual objects and the totality of objects. According to those making this objection, it is fine to ask about the reason for a particular object's existence. The problem supposedly arises when we ask about such reasons for groups of objects, or the collectivity of all objects, as in asking why the whole world exists.

The objectors do not doubt the fruitfulness of inquiring into the reasons for *individual objects*. However, once we have an explanation for each of various individual objects, they argue, there is no point in asking the further question: Why does this *collection of objects* exist? With the explanation for the individuals in hand, there is nothing left to be explained. Thus, it is illegitimate to demand an explanation for the world, as the collection or plurality of all things.

As Hume acknowledges, our everyday language routinely refers to groups of objects. He notes, for example, that we may speak of "several distinct countries" as a "kingdom," or of "several distinct members" as "one body."[3] However, Hume denies that the collection has an existence apart from the individual objects that comprise it.[4]

If collections of objects have no independent existence, then why do we speak as if they do? According to Hume, this habit is the result of nothing more than an "arbitrary act of the mind" that has no relation to the "actual nature of things." Since the collection has no independent reality, it requires no explanation apart from the explanation for each of the individual objects. As Hume writes, "Did I show you the particular causes of each individual in a collection of twenty particles of matter, I should think it very unreasonable, should you afterwards ask me, what was the cause of the

1 For example, some thinkers object to asking why there is a world on the grounds that there is no conceivable alternative. They argue that the idea of nothingness only makes sense in referring to the absence of particular things in a specified place, as when one says that there is nothing in the closet, indicating that the closet contains no umbrellas, jackets, shoes, or other such objects. By contrast, the argument goes, the idea of absolute nothingness is unintelligible and, therefore, so is asking why there is something rather than nothing. See Davies, *An Introduction to the Philosophy of Religion*, 60–61.

2 We have met Hume in two previous chapters. Chapter Three discusses his views on scientific natural laws and Chapter Six discusses his objections to teleological arguments for the existence of God.

3 Hume, *Dialogues concerning Natural Religion*, 190.

4 Hume, *Dialogues concerning Natural Religion*, 190. See also Paul Edwards, "The Cosmological Argument," in *The Cosmological Arguments: A Spectrum of Opinion*, ed. Donald R. Burrill (Anchor Books, 1967), 113.

whole twenty. This is sufficiently explained in explaining the cause of the parts."[5] If we have an explanation for each item in a travel bag—say, a towel, hairbrush, and bar of soap—there is no sense in asking additionally for an explanation of *all* the items in the bag. Philosopher Paul Edwards (1923–2004) expresses the same idea in saying, "If we have explained the individual members there is nothing additional left to be explained." Echoing Hume, he says that the demand for an explanation of the whole results from an "erroneous assumption that the series is something over and above the members of which it is composed."[6]

Based on their view that the whole requires no independent explanation, Hume and Edwards argue that there is no point in asking why the world exists. While people may speak of the "world" as the totality of all objects, Hume and Edwards deny that the world has a reality apart from each of the individual objects. So long as we can explain the individual objects, we have no reason to expect an additional explanation for the whole.

We noted that debate over the legitimacy of asking why the world exists implicates debates about what counts as an explanation. One example concerns whether an infinite regress can count as an explanation. The issue arises because some thinkers contend that there *must* be an explanation for the world as a whole—above and beyond explanations for the individual things comprising the world—because the only alternative is an infinite regress of individual causes, which cannot count as a satisfactory explanation.[7] However, Hume and Edwards deny that there is any problem with a series of individual explanations giving rise to an infinite regress. In their view, the only requirement is that each individual object has an explanation. If each individual object has an explanation for its existence, then we have no basis for demanding more. As Hume writes, in a chain "or succession of objects each part is caused by that which preceded it, and causes that which succeeds it. Where then is the difficulty?"[8]

In a radio debate with the Jesuit priest and philosopher Frederick Copleston (1907–94), philosopher Bertrand Russell (1872–1970) made

5 Hume, *Dialogues concerning Natural Religion*, 190–91. As noted earlier, Hume presents his arguments in the form of a dialogue among three characters: Cleanthes, Demea, and Philo. See Chapter Six. In this instance, the quoted language is expressed by Cleanthes.

6 Edwards, "The Cosmological Argument," 113.

7 Thomas Aquinas is among the most prominent examples. See Chapter Six. Much discussion over the legitimacy of asking why the world exists has taken place in the context of debates over cosmological arguments seeking to demonstrate God's existence. Aquinas, for example, uses the need to explain certain of the world's basic features—such as the fact that it contains things in motion—to motivate versions of his cosmological argument. Thinkers like Hume and Edwards have largely objected to the need for an explanation of the world's existence in the course of arguing that cosmological arguments are not successful in demonstrating God's existence. Nevertheless, their arguments have significance for the legitimacy of seeking to explain the world's existence more broadly.

8 Hume, *Dialogues concerning Natural Religion*, 190.

arguments resonating with those advanced by Hume over a century and a half earlier.[9] During the debate, Copleston and Russell disagree over the legitimacy of seeking an explanation for the "existence of ... the whole universe."[10] Copleston begins by making a number of points that have long played an important role in the thought of those defending the legitimacy of asking why the world exists. Rehearsing elements of Leibniz's cosmological argument,[11] Copleston observes that the world consists of contingent things (meaning that their existence depends on something else outside themselves). Since no individual objects can explain their own existence, Copleston argues, nothing in the world can explain the existence of the world as a whole. Against the view of Hume and Edwards, Copleston maintains that an infinite series of contingent causes cannot count as an explanation for the existence of the whole series. Illustrating the point, he writes, "If you add up chocolates you get chocolates after all and not a sheep.... So if you add up contingent beings to infinity, you still get contingent beings.... An infinite series of contingent beings will be ... as unable to cause itself as one contingent being."[12] Copleston concludes that only a being that is both necessary and external to the world can explain the world's existence and, therefore, such a being (God) must exist.[13]

In arguing against the legitimacy of asking why the world exists, Russell notes that we derive the very idea of a cause from observing the behavior of particular objects. However, we do not have the ability to observe the totality of all objects. Consequently, he argues, "the concept of cause is not applicable to the total."[14] The critical error of the inquiry regarding the world's existence lies in the fallacy of assuming that the totality of all things must have a cause or explanation just because individual things do. Countering Copleston's colorful analogy with one of his own, Russell writes, "Every man who exists has a mother," but it does not follow "that therefore the human race must have a mother.... [T]hat's a different logical sphere."[15] Russell's objection to the idea of a totality as something distinct from individual objects is so thoroughgoing that he asserts that the word "universe" lacks any meaning.[16]

Russell also rejects Copleston's reliance on the distinction between contingent and necessary existence. For Russell, the idea of necessity applies only to propositions. When we say that a proposition is necessary, we mean

9 Bertrand Russell, *Why I Am Not a Christian: And Other Essays on Religion and Related Subjects* (Routledge, 1957), 127–38.

10 Russell, *Why I Am Not a Christian*, 134.

11 See Chapter Six for a discussion of Leibniz's cosmological argument.

12 Russell, *Why I Am Not a Christian*, 127–38.

13 Russell, *Why I Am Not a Christian*, 127–38.

14 Russell, *Why I Am Not a Christian*, 134.

15 Russell, *Why I Am Not a Christian*, 135.

16 Russell, *Why I Am Not a Christian*, 132.

that its denial would entail a self-contradiction.[17] For example, it is necessarily true that a thing is identical to itself, because denying this would entail the self-contradiction that a thing both is and is not itself. The idea of necessity has no application to existence because it makes no sense to speak of an object's non-existence as entailing a self-contradiction. Thus, Russell does not merely deny that there are any necessary beings; rather, he denies that the idea of a necessary being even has any meaning.[18] Russell tells Copleston that by futilely seeking a necessary being to explain contingently existing objects, he "is looking for something which can't be got, and which one ought not to expect to get." Indeed, since there is no meaningful alternative, Russell says, we do not even need the term "contingent" to characterize the reasons why things exist. It is always the case that "the explanation of one thing is another which makes the other thing dependent on yet another...."[19] Ultimately, Russell concludes that there can be no explanation for the collection of all things and that the entire question is meaningless. He declares that "the notion of the world having an explanation is a mistake.... [It] is just there, and that's all."[20]

While Russell repudiates the very idea of necessary existence, one need not go that far to reject Copleston's argument for believing that the world must have an explanation. Edwards, for instance, argues that even granting the idea of necessary existence, we do not need to postulate any necessary beings to make explanations intelligible. Like Hume and Russell, he maintains that we can understand the causes of events and the objects around us through an ongoing process of explanation for contingent things. There is no need to trace all existing things to a "first cause" or necessary being.[21]

Many thinkers who reject the question of why the world exists share certain ideas about existence and explanation. What exists are individual objects. Collections have no existence independent of the individuals and, therefore, we can only expect explanations of the individuals. Imaginations to the contrary are mistaken habits of thought with no bearing on reality. Since objects depend on other objects for their being, explanations for one object refer to other objects, and there is no problem with an infinite series of such explanations. As long as each object has an explanation, we can expect no more.

These arguments fit congenially with an emphasis on empirical observation as the basis of explanation. We observe individual objects—and their interactions with other objects—but we cannot step out of our vantage

17 Russell, *Why I Am Not a Christian*, 127.
18 Russell, *Why I Am Not a Christian*, 129.
19 Russell, *Why I Am Not a Christian*, 132.
20 Russell, *Why I Am Not a Christian*, 134.
21 Edwards, "The Cosmological Argument," 118–20.

point to observe the totality of objects. Some advocates of these views contend that only the empirical methods of science can provide us with reliable knowledge about the things around us.[22] If, as the arguments discussed maintain, we only have a basis for explaining individual objects, then it is pointless to ask why the world exists, as if we could expect an explanation for something other than the individuals themselves.

As with the thinkers who oppose asking why the world exists, those defending the question have advanced a variety of arguments. In light of the themes commonly pursued by thinkers like Hume, Edwards, and Russell, it is not surprising that their opponents have also focused on basic issues regarding existence and explanation. One way to defend the question is to insist that collections or totalities of objects have an existence beyond that of the individual objects comprising them. That position enables us to think of the whole world as having an independent reality and, thus, as having a reason for existence that is not reducible to individual objects. As one scholar writes, "[I]f the universe is something more than the sum of its material parts, then an explanation of the existence of each material object in the universe would not be an explanation of the universe."[23]

Relatedly, one may raise questions about what counts as an individual object. The case against asking why the world exists gives central importance to the idea of an "individual object" by claiming that we can only legitimately seek explanations for individual objects (and not for collections of individual objects). The phrase "individual object" may seem straightforward enough at first as examples like a chair, pebble, or refrigerator, come to mind. On reflection, however, it is not obvious what makes something an individual object rather than a part or combination of other individual objects. If I buy a replacement drawer for my refrigerator, is the drawer an individual object? If the drawer has a detachable handle, is the handle an individual object or merely part of the drawer? We may ultimately wonder if each subatomic particle comprising a chair, pebble, or refrigerator is an individual object. Moving to the other extreme in terms of scale, we might view the entire universe as an individual object. For example, defining the universe as "the physical object consisting of all physical objects including the Earth spatially related to each other and to no other physical object" uniquely identifies the universe as a distinct object.[24] To the extent that the case against asking why the world exists depends on the idea that only

22 Russell, for example, stresses scientific causation as the basis for our knowledge and explanations of events. Richard M. Gale, *On the Nature and Existence of God* (Cambridge University Press, 1991), 215–16. See Russell, *Why I Am Not a Christian*, 132.

23 Joseph K. Campbell, "Hume's Refutation of the Cosmological Argument," *International Journal for Philosophy of Religion* 40, no. 3 (1996): 166–67.

24 Swinburne, *The Existence of God*, 134.

individual objects can be explained, it owes us a workable definition of the term that rules out the "universe being an individual in its own right."[25]

It might initially seem odd to consider the entire universe as an individual, rather than a collection of objects.[26] But many philosophers think there is good reason for doing so. Here is an example of reasoning that might lead us to consider the universe as an individual. Think about all the parts of a bike. There are handlebars, tires, brakes, pedals, etc. But, in addition to all of these individual parts, there is another thing: the bike. We might say that the individual parts of a bike *fuse* together to form the object that is the bike itself. Suppose you were to remove one of the tires from a bike and replace it with a new one. Does this make the bike no longer the same bike? If not, this suggests that the bike is not merely a composite of parts; the bike itself is an individual thing as well. The same considerations that apply to the bike can apply to a wide range of other things. For example, if we can say that the bike exists because its parts are put together, then why should we not also say that the fusion of, say, my left shoe and the sidewalk underneath it are an individual object? After all, it, too, has parts and is put together in some way. Examples like these lead some philosophers to accept an *unrestricted principle of fusion*, which holds that the fusion of any collection of individual objects constitutes a new individual object. Applying his principle of fusion to all of the objects in the world generates the individual that is the fusion of all objects: the world itself.

One strategy for defending the question of why the world exists, then, is to show that the universe has an independent existence (either as a collection of individual objects or as itself being an individual object). However, defending the question does not necessarily depend on that claim. One may allow that collections of individual objects have no independent existence over and above the individuals which make them up, but nevertheless argue that the world has an explanation apart from the explanations for the individual objects composing it.[27] The key claim in this kind of argument is that certain explanations apply to collections of objects that do not come into play when examining the individual objects in isolation.[28]

An example of an explanation that might apply only at the level of the collection is one concerning the reason that different individual objects

25 Gale, *On the Nature and Existence of God*, 215.

26 Broadly speaking, the idea that the world is one thing traces to some of the ancient philosophers. As discussed earlier, for example, Parmenides conceived of the world as one single being. See Chapter Two.

27 Copleston, for example, does "not say that the universe is something different from the objects which compose it," but nevertheless argues that the universe has an explanation—based in necessity—which does not apply to any of the contingently existing individual objects. Russell, *Why I Am Not a Christian*, 133.

28 See Campbell, "Hume's Refutation of the Cosmological Argument," 165.

became aggregated. One scholar offers the example of a heap of rocks. We can imagine that the swinging of a worker's sledgehammer explains the formation of each rock in the heap but that only the assembly of the rocks into a heap by another worker explains the heap itself.[29] Or suppose that three people each take a different gift to a party. The explanation for why each gift is at the party is that a particular person brought it. However, the explanation for why all three gifts are at the party might be distinct, such as that a fourth person asked each of the others to take a gift.[30]

Considering a collection may also bring new explanations into view with respect to the manner in which objects are arranged together.[31] Individual items needed for the construction of a computer may have separate explanations in terms of the factory mechanisms that produced them. However, none of those explanations at the individual level would reveal how the items came together in the form of a functioning machine. For that, we must refer to a distinct process that configured the separate parts into a computer. In that sense, the computer has an explanation that does not reduce to the explanations of the individual items comprising it.[32]

Another reason that new explanations may come into play at the level of the collection is that different objects often have common qualities. Suppose that we have an explanation for a number of individual objects that share a certain quality. There might nevertheless be another kind of explanation that accounts for why there are any objects at all with that quality. For example, we might explain the existence of any given frog, toad, or salamander in terms of the reproductive process that generated it. Yet, we can also ask why there are any animals at all possessing the ability to live in both water and land. The explanation for the existence of amphibious animals might turn on larger forces not evident at the level of the individual organisms. On a grander scale, we might learn something through investigation into the origins of life that would not be captured through inquiry into the explanation for any individual living thing.

The line of argument just described suggests a potential response to one of Russell's examples. To support his view that explanations only apply to individual objects, Russell says, "Every man who exists has a mother … but obviously the human race hasn't a mother—that's a different logical sphere."[33] Russell here assumes that the explanation for why there are any

29 Gale, *On the Nature and Existence of God*, 217.

30 A similar example is described in Campbell, "Hume's Refutation of the Cosmological Argument," 165.

31 Bruce Reichenbach, "Cosmological Argument," in *The Stanford Encyclopedia of Philosophy*, ed. Edward N. Zalta and Uri Nodelman (2023), §§ 7.2–7.3, accessed December 10, 2024, https://plato.stanford.edu/archives/win2023/entries/cosmological-argument/.

32 Gale, *On the Nature and Existence of God*, 217.

33 Russell, *Why I Am Not a Christian*, 135.

human beings at all must take the same form—a particular parent—as the explanation for each individual human being. However, another possibility is that explaining the existence of each individual human being leaves something out that has not yet been explained: why there are any human beings at all. Answering *that* question might require reliance on reasons that do not apply at the level of the individuals.

Similar reasoning can be applied to the question of why there is a physical world. Thinkers like Hume, Edwards, and Russell contend that once we explain the existence of each individual object in the world there is nothing left to be explained. However, the individual objects share certain qualities, such as occupying space, persisting over time, and causally influencing other things in space and time. Thus, there is still something left unexplained: why there are any objects at all possessing those qualities. And answering *that* question may require relying on reasons that do not apply at the level of the individuals.

We have seen that the debate over asking why the world exists implicates issues regarding what can count as explanations. Thinkers who deny the legitimacy of asking why the world exists often have a more restrictive view of explanation. Some limit explanation to the kinds of causal explanations associated with empirical science. From that vantage point, the world cannot have an explanation that is fundamentally distinct in character from the explanations for individual objects. Each individual object finds explanation in other individual objects, and if this leads to an infinite regress, then so be it.

By contrast, defenders of asking why the world exists commonly allow room for explanations that do not take the form of scientific theories or causal mechanisms. For example, Copleston—echoing Leibniz—contends that the explanations for individual objects are contingent, but that individual objects cannot explain why there are any contingent beings in the first place. The explanation for the whole world of contingent objects is fundamentally different: a necessary (self-explanatory) being. Another example is John Leslie's view that the explanation for why any physical objects exist is goodness itself, where goodness is not physical and does not operate according to the causal mechanisms described by science.[34] The ultimate explanation for the existence of individual objects transcends the individual objects themselves.

The debate over the legitimacy of asking why the world exists is connected with other basic issues. There is, of course, a good deal of variety among thinkers on either side of that debate. However, there are also common threads, with the opposing positions often turning on what we recognize as real and what counts as a satisfying explanation.

---

34 See Chapter Five.

Nāgārjuna's *Madhyamaka* Buddhism

In this and the next section, we consider schools of thought centered on views that render the puzzle of existence incoherent or beside the point. We begin with Nāgārjuna's *Madhyamaka* Buddhism and then examine Śaṅkara's Advaita Vedānta Hinduism.

Nāgārjuna is a Buddhist monk, philosopher, and mystic, whose work established an important school of *Madhyamaka* Buddhism. While historians are not certain of the details, they believe Nāgārjuna was born in South India around 150 CE, a period of substantial debate among monks over the Buddha's teachings.[35] Of Nāgārjuna's writings—ranging from commentaries on existing texts to advice for rulers—the best-known is his treatise discussing the Buddha's ideas, *Mūlamadhyamakakārikā* ("Fundamental Verses on the Middle Way"), which became the basis of *Madhyamaka* Buddhism (meaning the "Middle Way").[36] While there are many versions of Buddhism, we will focus on Nāgārjuna's ideas, which have been influential in Indian philosophy and Buddhism around the world.[37] Many consider him the most significant commentator on Buddhist thought next to the Buddha himself.[38]

In our exploration of the ultimate question, we have encountered prominent theories that identify something with an independent existence that is supposed to explain why there is a world. Axiological theories, for example, claim that goodness—with an existence independent of anything else—is responsible for the world, and theistic answers accord a similar role to a divine being. As we will see, by denying that anything has an independent existence, Nāgārjuna's thought rules out any attempt to explain the world's existence in this fashion. At the same time, Nāgārjuna's viewpoint subverts attempts to answer the puzzle of existence in an even more far-reaching manner by declining to recognize the question as meaningful in the first place.

A core idea in Nāgārjuna's thought—sometimes referred to as the "doctrine of dependent origination"—is that all things are interdependent. According to this principle, nothing exists independently of other things or has a fixed essence of its own.[39] Nāgārjuna's philosophy centers on his

35 Bina Gupta, *An Introduction to Indian Philosophies: Perspectives on Reality, Knowledge, and Freedom* (Routledge, 2012), 211.

36 Richard King, *Indian Philosophy: An Introduction to Hindu and Buddhist Thought* (Georgetown University Press, 1999), 91, 120; Jay Garfield, *The Fundamental Wisdom of the Middle Way: Nagarjuna's Mulamadhyamakakarika* (Oxford University Press, 1995), 87.

37 Jan Christoph Westerhoff, "Nāgārjuna," in *The Stanford Encyclopedia of Philosophy*, ed. Edward N. Zalta and Uri Nodelman (2024), introductory section, accessed December 18, 2024, https://plato.stanford.edu/archives/sum2024/entries/nagarjuna/.

38 Jan Westerhoff, *Nāgārjuna's Madhyamaka: A Philosophical Introduction* (Oxford University Press, 2009), 4.

39 Gupta, *An Introduction to Indian Philosophy*, 211.

claim that the world is empty of *svabhāva*.[40] We will examine the meaning of *svabhāva*, why *Nāgārjuna* denies its existence, and implications of that position.

While English has no equivalent, common translations of *svabhāva* include "intrinsic nature," "inherent existence," "essence," and "own-being."[41] As Jan Westerhoff writes, having *svabhāva* would mean that something "exists objectively" with qualities that are "independent of other objects, human concepts, or interests."[42] An object with *svabhāva* would not depend on anything else, indicating that its existence was not the effect or product of any other things.[43] Such an object could "be part of the basic furniture of the world," or the "ontological rock-bottom" on which other things ultimately rest."[44] In denying that anything has *svabhāva*, Nāgārjuna proclaims that everything is interdependent. Nothing arises—or could arise—on its own.[45]

Nāgārjuna's case against *svabhāva* is ambitious. He does not merely suggest reasons for rejecting *svabhāva*, but, rather, claims that the very idea of *svabhāva* is incoherent. Accepting the reality of *svabhāva*, Nāgārjuna contends, is inconsistent with aspects of our experience, such as causation, motion, and change.[46] We will focus principally on his arguments concerning causation, which illustrate a strategy he uses repeatedly to refute the existence of *svabhāva*.

Specifically, Nāgārjuna employs a *reductio ad absurdum* argument (or *reductio*, for short) to show that accepting the reality of *svabhāva* entails the unacceptable conclusion that causation is impossible.[47] The *reductio* begins by assuming that a cause and its effect both have their own *svabhāva*. On this assumption, the cause and the effect can each exist independently of the other.[48] To elaborate the *reductio*, Nāgārjuna claims that there are only four possible bases of causation and argues that all of them are inconsistent with the existence of *svabhāva*. The four possibilities are (1) that an effect is caused by itself, (2) that an effect is caused by something other than itself, (3) that an effect is caused both by itself and something else, and (4) that an effect is not caused either by itself or something else.[49]

---

40 Westerhoff, "Nāgārjuna," § 2.

41 Westerhoff, *Nāgārjuna's Madhyamaka*, 19–20.

42 Westerhoff, *Nāgārjuna's Madhyamaka*, 199.

43 Westerhoff, *Nāgārjuna's Madhyamaka*, 24; Garfield, *The Fundamental Wisdom of the Middle Way*, 221.

44 Westerhoff, "Nāgārjuna," § 2.

45 Garfield, *The Fundamental Wisdom of the Middle Way*, 221.

46 Garfield, *The Fundamental Wisdom of the Middle Way*, 88–89.

47 As discussed in Chapter Six, a *reductio ad absurdum* argument assumes the opposite of the conclusion that the thinker wishes to support and then seeks to show that the assumption generates absurd results. The idea is that if assuming the truth of X leads to something obviously false—such as a logical contradiction—then X must itself be false.

48 Westerhoff, "Nāgārjuna," § 3.1.

49 Westerhoff, *Nāgārjuna's Madhyamaka*, 14.

Regarding the first possibility, Nāgārjuna identifies two senses in which we might say that an effect causes itself. The first is that the cause and the effect are identical to one another (so there is no distinction between them in the first place).[50] Some thinkers assert, for instance, that the existence of an object at one moment is the cause of that same object's existence in the next moment. From this standpoint, it is self-causation that explains the persistence of objects through time. Nāgārjuna, however, finds two fatal flaws in this view. One problem is that, as our experience attests, causes precede effects, yet it is incoherent to say that an object precedes (or exists before) itself.[51] Another difficulty is that causes are plainly not identical to their effects. For example, a lit match might bring about an explosion, but this does not mean that the match and the explosion are the same thing.[52]

A second sense in which an effect might cause itself is that the cause already contains the effect within itself.[53] Nāgārjuna also denies that this kind of self-causation is viable. For one thing, if a cause already includes the effect, then there is no need for the operation of the cause to bring about the effect; it is already there.[54] Additionally, we know that causes can be present without their effects. We can observe the lit match without the resulting explosion.[55] To be sure, Nāgārjuna recognizes that we can often predict an effect from observing the cause, but this is not the same as saying that the effect is already present.[56]

Having dismissed self-causation, Nāgārjuna considers the second possibility: that an effect is brought about by something other than itself. He contends that this possibility violates the *reductio's* assumption that cause and effect both have *svabhāva*. The problem is that the very idea of a causal relation entails a relationship of dependency between the cause and effect. Without any such dependency between two things, one cannot cause the other. It makes no sense to say that an effect could exist without its cause. The reverse is also true: a cause cannot exist without its effect, since it would not count as a "cause" without the effect whose occurrence is what qualifies it as a cause. If the cause and effect are interdependent—as must be the case—then, by definition, they do not have *svabhāva*.[57]

With the first two possibilities dismissed, Nāgārjuna finds little difficulty in excluding the other two. If it makes no sense to say that an effect is caused

50 Westerhoff, *Nāgārjuna's Madhyamaka*, 99–100.
51 Westerhoff, "Nāgārjuna," § 3.1.1.
52 Westerhoff, *Nāgārjuna's Madhyamaka*, 100.
53 Westerhoff, *Nāgārjuna's Madhyamaka*, 99–100.
54 Westerhoff, *Nāgārjuna's Madhyamaka*, 102.
55 Westerhoff, "Nāgārjuna," § 3.1.1.
56 Westerhoff, *Nāgārjuna's Madhyamaka*, 102.
57 Westerhoff, "Nāgārjuna," § 3.1.2.

either by itself or by something else, then claiming that an effect results from both—the third possibility—is no more cogent.[58] The last possibility—that an effect is brought about neither by itself nor by something else would mean that there simply was no causation at all.[59] Without causation, however, we would be unable to account for obvious aspects of our experience. The world that we experience every day is causally ordered, but a world without causation would be one of nothing but random occurrences with no connection between one event and the next.[60] In addition, it is obvious that we learn about the things around us from observation, but this requires some kind of causal interaction between the world and our sensory perceptions.[61]

The conclusion of Nāgārjuna's *reductio* is not that we should deny the reality of causation. Rather, the point is that causation is inconsistent with *svabhāva*. Since it is absurd to deny causation, we must reject *svabhāva*. This means that a cause and its effect do not exist as independent, self-sufficient entities with their own essential natures.[62] Instead, causes and effects are mutually interdependent.[63] Nāgārjuna's larger claim is that *everything* originates in a way that is dependent on other things. The wooden cabinet is dependent on the carpenter who built it from wood, which is dependent on the work of the logger, which is dependent on the tree, which is dependent on the seed from which it grew, and so on.[64]

Nāgārjuna uses similar arguments to show that the reality of *svabhāva* is inconsistent with other obvious aspects of our experience, such as motion, change, and objects having essential properties.[65] Regarding the last, Nāgārjuna notes our tendency to view the world as comprising separate objects with their own essential and distinct properties. People typically take it for granted that individual objects exist independently of their particular properties. Nāgārjuna contends, though, that the idea of objects and their properties possessing *svabhāva* does not hold up to scrutiny, since they are inevitably dependent on one another. The idea of an object without any properties is unintelligible. Conversely, we cannot conceive of properties apart from the existence of any objects that might instantiate them. In sum, without properties, there could be no objects to exhibit them, and, without objects, there could be no properties that they might exhibit.[66]

58 Gupta, *An Introduction to Indian Philosophies*, 212.
59 Westerhoff, "Nāgārjuna," § 3.1.4.
60 Sue Hamilton, *Indian Philosophy: A Very Short Introduction* (Oxford University Press, 2001), 95.
61 Westerhoff, *Nāgārjuna's Madhyamaka*, 112.
62 Westerhoff, *Nāgārjuna's Madhyamaka*, 93–94.
63 Gupta, *An Introduction to Indian Philosophies*, 213
64 King, *Indian Philosophy*, 120.
65 Westerhoff, "Nāgārjuna," § 3.2.
66 Westerhoff, *Nāgārjuna's Madhyamaka*, 112.

Furthermore, Nāgārjuna maintains, the notion that objects have essential properties rests on a failure to recognize the arbitrariness of our concepts. Consider a blue vase. We reflexively think of the vase as the object and the blue coloring as one of its properties. This assumption underlies our conventional thought that the vase is defined by its essential properties, such as having the characteristic shape of a vase. As an independently existing object, we assume, the vase may take on any number of distinct properties, such as being blue (or another color), or heavy, or wooden. However, Nāgārjuna notes, selecting shape as the essential property that makes the object what it is—a vase—rather than some other object, is merely a convention. Instead of describing a vase-shaped object as manifesting the property of blue-ness, we could, with no less warrant, say that an essentially blue object manifests the property of vase-shaped-ness. Apart from our arbitrary determinations, there is nothing that marks out one particular property as the one that essentially defines an object. In the end, Nāgārjuna concludes, the only proper way to make sense of the world is to recognize that everything in it—including what we normally think of as individual objects and properties—is mutually interdependent.[67]

Nāgārjuna's position is far-reaching. He claims that nothing possesses *svabhāva*. There is nothing that has an essential, intrinsic, independent, or enduring nature. Why, then, does the idea that things have *svabhāva* get such a hold on us? We are wont to regard the world as comprising individual objects with their own distinct origin stories, properties, and essential natures. When we witness a match lighting a candle, we consider the match, the candle, and the resulting flame as self-sufficient objects with their own independent existences. We apply this habit of thought not only to the objects around us, but also to our *selves*. We presume the existence of a self that is distinct from the flood of sensations, feelings, and thoughts that continuously washes over us. Although most do not know the term, people in effect imagine that they have a self which possesses *svabhāva*.

According to Nāgārjuna, the reason for all of these mistaken impressions regarding *svabhāva* is that we superimpose certain inaccurate ideas onto objects. These superimpositions are not merely the error of rogue individuals on particular instances. The dynamic in play runs deeper. Our pervasive errors are the result of ingrained cognitive defaults. Without making conscious determinations, we routinely make various assumptions in an effort to make sense of the world.[68] The common practices by which we superimpose myriad concepts and terms onto the world are merely human conventions. Nāgārjuna's claim is not that the things our language refers to

67 Westerhoff, "Nāgārjuna," § 3.2.

68 Westerhoff, *Nāgārjuna's Madhyamaka*, 13.

do not exist. Rather, his crucial position is that all of these things exist only in a mutually interdependent manner subject to our superimposition of conceptual frameworks.[69] Notwithstanding our shared tendencies, there is in fact no world that exists entirely independent of human concerns and interests. It is not possible to coherently speak of the world as it *really* is without us in the picture at all.[70]

To explain how it is that we can regularly communicate on the basis of shared (false) assumptions, Nāgārjuna distinguishes between conventional and absolute truth. Conventional truth hinges on the particular conceptual and linguistic systems that people develop in navigating the world.[71] If I speak of a popsicle as being really cold and sweet—as if it were an independently existing thing with its own distinct and essential properties—my description may be true, but only at the level of conventional truth, which is pragmatic and relative to our language.[72] In addition to conventional truth, however, there is also a higher level of truth: the absolute truth. At this higher level, the ultimate, absolute truth is that nothing has *svabhāva*, and our impressions to the contrary are merely misleading appearances.[73]

Conventional and absolute truth do not represent fundamentally different realms of being. They are two levels of understanding or meaning.[74] As one scholar summarizes the distinction, Nāgārjuna "regards individuals and properties as linguistically or conceptually mediated projections of at best pragmatic importance, but not as objective features of a mind-independent reality."[75] In characterizing the world as empty of *svabhāva*, Nāgārjuna does not claim that emptiness is an entity in itself, or that it is the essence of all things. That would be to simply substitute one version of *svabhāva* for another.[76] He does not equate reality with nothingness. Nāgārjuna views the distinction between absolute and conventional truth as enabling him to reject *svabhāva* without falling into nihilism. The crucial claim is that all things are mutually relative and interdependent.[77] Thus, Nāgārjuna adopts a "middle way" regarding the world as we commonly (and mistakenly) perceive it (that is, as comprising many different independent things with distinct properties). He does not affirm that the world as we perceive it really exists, because, as we have seen, this contradicts the absolute truth that there

69 Westerhoff, "Nāgārjuna," § 3.5.

70 Westerhoff, "Nāgārjuna," § 3.5.

71 Gupta, *An Introduction to Indian Philosophies*, 215.

72 Peter Harvey, *An Introduction to Buddhism: Teachings, History and Practices* (Cambridge University Press, 1990), 99.

73 King, *Indian Philosophy*, 125.

74 King, *Indian Philosophy*, 124.

75 Westerhoff, *Nāgārjuna's Madhyamaka*, 15–16.

76 Garfield, *The Fundamental Wisdom of the Middle Way*, 90–91.

77 King, *Indian Philosophy*, 123.

is no *svabhāva*. Yet he also does not state that the world as it appears to us does not exist in any sense at all, since we do in fact perceive the world as filled with independently existing individual things.[78]

Nāgārjuna's distinction between conventional and absolute truth makes it critical that we pay attention to the intentions of the various statements that we make about the world. Any assertions that we make referring to individual things and properties as if they had *svabhāva* will be false if intended as statements about the ultimate or absolute truth. This follows from Nāgārjuna's claims about emptiness and dependent origination. On the other hand, such assertions may be true in a dependent or relative sense if understood as pertaining only to phenomena as we experience and talk about them in conventional terms.[79]

As is often the case in Buddhist thought, Nāgārjuna is interested in far more than making correct claims about the nature of reality. His metaphysical claims—about the nature of what exists—have vital implications for how people should conduct their lives, especially with respect to the minimization of suffering. A great deal of anguish, Nāgārjuna maintains, results from assuming the reality of *svabhāva*. Imagining things as having essential, enduring natures, people form emotional attachments to them in a way that proves painful.[80] Since we ourselves superimpose *svabhāva* on things that lack it, it is possible to eliminate the falsehoods. That does not mean that this will be easy. Since our superimposition of *svabhāva* is so ingrained, it is challenging to suppress. Merely presenting someone with the absolute truth that *svabhāva* is unreal will not suffice. Instead, we must root out our ingrained tendencies through sustained exercises and practices. With proper training, one may even come to see and interact with the world in a fundamentally different—and more salutary—manner.[81]

Nāgārjuna's claims have radical metaphysical implications, overturning views that are often taken for granted. Taking a cue from modern physics, many people assume the reality of certain fundamental entities—such as electrons and quarks—that do not depend on anything more elementary for their existence. By interacting and combining with other fundamental entities, these subatomic particles are supposed to serve as the building blocks of other things. Underlying these common beliefs is the recognition of existential bedrock.[82] We also tend to assume that it is possible to identify independent objects with defining characteristics whose properties change

78 Hamilton, *Indian Philosophy*, 96; Harvey, *An Introduction to Buddhism*.
79 Garfield, *The Fundamental Wisdom of the Middle Way*, 90.
80 Westerhoff, "Nāgārjuna," § 2.
81 Westerhoff, *Nāgārjuna's Madhyamaka*, 13.
82 Westerhoff, *Nāgārjuna's Madhyamaka*, 199–200.

over time while their essential nature persists. However, Nāgārjuna's insistence on mutual interdependence is so sweeping that it undermines categories that are practically part of the contemporary observer's pre-loaded package of common understandings. Distinctions jettisoned by Nāgārjuna's far-reaching claims include, for instance, those between an object and its attributes, a whole and its components, and universals and particulars.[83]

The counterintuitive implications of Nāgārjuna's philosophy also concern our selves. Each of us has a stream of sensations, feelings, thoughts, opinions, preferences, and the like. In addition, each of us seems to have a personality or personal identity with particular traits and characteristics. We commonly assume that underlying all of these experiences and properties is something that unifies them so that they all pertain to the same person. There is a self with an essential nature that instantiates my various properties and makes all of my actions those of the same individual agent.[84] However, from Nāgārjuna's standpoint, that assumption is wrong because it views the self as an independently existing thing. His rejection of *svabhāva* applies no less to our selves than it does to everything else. Just as we have no basis for distinguishing properties that are essential from the rest, we have no basis for picking out anything in particular from our stream of consciousness as the essential self. What we are tempted to call a "self" is not an independently existing thing in its own right, but merely a set of interconnected mental and physical events.[85]

The centering of Nāgārjuna's philosophy on emptiness—with its wholesale rejection of essential natures and across-the-broad insistence on the mutual interdependence of all things—has vital implications for the puzzle of existence. Nāgārjuna rejects the very idea that anything has an independent existence or its own nature. He does not even identify emptiness as the essential nature of anything. There is not something out there—Emptiness—that counts as the ultimate reality or nature of all things. In that sense, emptiness, too, is empty. Thus, Nāgārjuna is not making a claim about "absolute nothingness" as the actual state of affairs. Rather, the overriding import of emptiness is that there are no exceptions to the doctrine of dependent origination. There is not anything that is independent of other things. The upshot of Nāgārjuna's philosophy is that we cannot theorize about the nature of things, full stop. As Richard King writes, the "language of ontology (what exists and does not exist) must be abandoned since both existence and nonexistence both miss the mark...."[86] However, if we cannot

83 Gupta, *An Introduction to Indian Philosophy*, 214.
84 Westerhoff, *Nāgārjuna's Madhyamaka*, 16, 159, 208.
85 Westerhoff, *Nāgārjuna's Madhyamaka*, 16, 159, 208.
86 King, *Indian Philosophy*, 121.

talk about existence, then, as Jay Garfield, notes, we must abandon "metaphysical doctrines regarding the nature of things" altogether.[87]

In previous chapters, we have seen prominent theories that answer the puzzle of existence by identifying something with an independent existence. For example, axiological theorists identify goodness as having an independent reality that is responsible for the world's existence. Theistic answers have a broadly similar structure in claiming that an independently existing God created the world. Given Nāgārjuna's denial that anything has an independent existence, his philosophy rules out any attempt of that kind to explain why the world exists.

Furthermore, Nāgārjuna's thought subverts attempts to answer the puzzle of existence in an even more far-reaching manner. By turning away from ontological theorizing about existence across the board, Nāgārjuna's viewpoint suggests a quietist stance toward the puzzle of existence. His philosophy does not so much reject particular answers to the puzzle of existence as it simply refuses to recognize the question as salient, worthwhile, or intelligible in the first place.

Śaṅkara's Advaita Vedānta Hinduism

Hinduism is rooted in the Vedas, scriptural texts that include hymns, ritualistic practices, theology, and philosophical speculations, often presented as stories or conversations.[88] Schools of Vedānta (meaning the "end of the Veda") Hinduism concentrate on later portions of the Vedic literature, known as the Upaniṣads. Reflecting the character of the Upaniṣads themselves, Vedānta Hinduism places a greater relative emphasis than did earlier schools on philosophical ideas as compared to ritualistic traditions.[89] While there are many competing understandings, Vedānta is a particularly well-known branch of Hinduism, and Advaita is an especially prominent Vedānta school. So influential is Advaita Vedānta that many equate it with Vedānta generally,[90] with one scholar calling it "the most widely known system of Indian philosophy in the East and the West alike."[91]

87 Garfield, *The Fundamental Wisdom of the Middle Way*, 88–89.

88 Karl Potter, *The Encyclopedia of Indian Philosophies, Vol. 3: Advaita Vedanta Up to Samkara and His Pupils* (Princeton University Press, 2014), 3.

89 Jeaneane Fowler, *Perspectives on Reality: An Introduction to the Philosophy of Hinduism* (Sussex Academic Press, 2002), 238.

90 Potter, *The Encyclopedia of Indian Philosophies*, 6.

91 Gupta, *An Introduction to Indian Philosophy*, 225.

While there is little evidence beyond the numerous works he authored, historians believe that Śaṅkara was born in south India around 800 CE.[92] As a philosopher, teacher, and interpreter of significant texts, Śaṅkara was a pivotal figure in the development and systematization of ideas discussed in the Upaniṣads.[93] Faced with ambiguities and inconsistencies in the texts, Śaṅkara aimed to harmonize them within a coherent body of ideas.[94] He is not only the most revered authority on Advaita Vedānta, but is also one of the best-known Indian thinkers, with his thought influential in the shaping of Hinduism more generally.[95]

As we will examine, even though Śaṅkara did not himself focus on the topic, his thought has radical implications for the quest to explain the world's existence. Posing the puzzle of existence tends to take for granted that the world comprises many things—including different people—that interact with one another. By denying this basic assumption, Śaṅkara excludes any possibility of explaining why one thing exists by referring to some other thing. For Śaṅkara, there are no distinctions between different things. All is one. Thus, Śaṅkara's thought rules out anything like the more familiar answers to the puzzle of existence. More than this, his philosophy as a whole is simply not interested in addressing the problem. In the end, Śaṅkara aims to liberate us from suffering by exposing harmful illusions in our common ways of understanding the world. He is ultimately interested not so much in explaining the world as in explaining it away.

The central claim of Śaṅkara's Vedānta Advaita philosophy is that all being is one. Notwithstanding appearances, there are not really multiple, different entities.[96] ("Advaita" is from the Sanskrit term for "not-two" or "non-dual.)[97] The one being is called *brahman*. As King summarizes the idea, "Only *brahman* is."[98] The oneness of being means that *brahman* does not reduce to anything else, and nothing exists apart from it. It does not have parts, and it is not part of anything else. *Brahman* is the single foundation on which everything rests.[99] It contains no differentiation or relational properties.[100]

---

92 Neil Dalal, "Śaṅkara," in *The Stanford Encyclopedia of Philosophy*, ed. Edward N. Zalta (2021), § 1, accessed December 22, 2024, https://plato.stanford.edu/archives/win2021/entries/shankara/. Scholars estimate that he died at the age of 32. Gupta, *An Introduction to Indian Philosophy*, 118.

93 Dalal, "Śaṅkara," Introduction, § 1.

94 Fowler, *Perspectives on Reality*, 243–44.

95 King, *Indian Philosophy*, 214.

96 R.N. Dandekar, "Vedānta," in *Encyclopedia of Religion*, ed. Lindsay Jones (Macmillan Reference, 2005), 9546.

97 Anantanand Rambachan, *The Advaita Worldview: God, World, and Humanity* (State University of New York Press, 2006), 68.

98 King, *Indian Philosophy*, 216.

99 Dalal, "Śaṅkara," § 2; King, *Indian Philosophy*, 216.

100 Dalal, "Śaṅkara," § 3; Potter, *The Encyclopedia of Indian Philosophies*, 7.

Brahman's eternal, changeless essential character and lack of dependence on anything else mark it as real.[101] *Brahman*'s perfect, infinite nature makes it difficult to describe. Since defining something inherently means limiting it, we can only positively affirm the nature of a finite being. Our finite language cannot positively portray *brahman*, which is limitless. Even describing *brahman* as one is problematic, since we can only understand the meaning of numbers in relation to other numbers, yet, as noted, *brahman* defies differences and relations.[102] It is not that *brahman* is a blank space or empty void, but, rather, that we can most appropriately identify it by saying what it is *not*.[103] *Brahman* does *not* have a beginning or end, for instance, it does *not* change, and it is *not* subject to space, time, or causation.[104] As such, Śaṅkara concludes, the intrinsic nature of *brahman* is pure, self-illuminating, non-intentional consciousness. Given the unity and immutability of reality, *brahman* is not composed of—or reducible to—different individual consciousnesses that change over time.[105] Additionally, consciousness is not merely a quality of *brahman*, as this would suggest a distinction between *brahman* and one of its attributes.[106]

At the same time, our experiences give the impression of numerous different things around us that are continually changing. Our ongoing phenomenal experiences are at odds with Śaṅkara's description of reality. The dramatic gap between the nature of *brahman* and our impressions poses a challenge. Śaṅkara does not deny that there is such a gap. Nor does he necessarily purport to address every philosophical question or remove all mystery regarding *brahman*. Nevertheless, he develops an account that preserves *brahman* while acknowledging the character of our experiences. As noted, more than explaining the sum of our phenomenal experiences, he aims "to explain it away."[107]

Śaṅkara insists that his discussion of our experiences comports with his assertions that all is *brahman*, which never changes.[108] How, then, can his account accommodate what our experiences are like? Śaṅkara's attempt to reconcile *brahman* and the world relies on two major strategies: viewing reality from two different perspectives, and recognizing two different levels of existence. Since *brahman* is the ground of everything, the world that we experience

101 Dalal, "Śaṅkara," Introduction, § 2.1.

102 Rambachan, "The Advaita Worldview," 68.

103 Dandekar, "Vedānta," 9546.

104 Dalal, "Śaṅkara," § 2; Gupta, *An Introduction to Indian Philosophies*, 226.

105 Dalal, "Śaṅkara," Introduction, § 3.3.

106 Gupta, *An Introduction to Indian Philosophies*, 226.

107 Hans Torwesten, *Vedanta: Heart of Hinduism*, adapted by Lily Rossett, from a translation from the German by John Phillips (Grove Weidenfeld, 1991), 123.

108 King, *Indian Philosophy*, 213–17.

must be dependent on *brahman* and is an effect of it.[109] *Brahman* has a mysterious power to manifest as our phenomenal experiences. Our ignorance plays a significant role here, too, as it superimposes qualities like plurality and changeability onto *brahman*, which does not actually have those qualities.[110]

Śaṅkara illustrates the relation between *brahman* and the phenomenal world with an analogy. Consider pottery. The clay can take many forms at different times, from a shapeless lump to a mug or candleholder. Regardless of the form it takes, though, the clay itself remains the same. Similarly, *brahman* can take on many appearances without itself undergoing change.[111]

Another analogy expresses the significance of perspective. To the audience at a magic show, various things that are not real appear to occur. Yet the magician suffers no deception, perceiving things only as they really are. Like the magic show audience, we are deceived by our ignorance into perceiving *brahman* as plural and changing, but no such falsities appear from the perspective of *brahman* itself. The phenomenal world is a dependent effect of *brahman*, and has no reality apart from it. *Brahman* itself does not transform any more than a rope mistaken for a snake actually turns into a snake.[112]

The relation between *brahman* and the world is asymmetrical, with the latter existing only in a lesser sense than the former. The world has a lesser kind of existence because it derives from *brahman*, but *brahman* depends on nothing outside itself. As a clay pot has no being apart from the clay, the world has no being apart from *brahman*. The apparent multiplicity and changeability of things we perceive are illusory and not consistent with *brahman*'s enduring, unified nature. The concepts we ascribe to them are just names that do not accord with reality.[113]

Nevertheless, Śaṅkara does not relegate the world to a status of absolute non-existence. Instead, he accords it a middle status between non-existence and the full, ultimate existence that *brahman* enjoys. To convey the world's intermediate ontological place, Śaṅkara contrasts it with examples of things with an even lower status, such as hallucinations, mirages, delusions, misperceptions, and self-contradictions (like the idea of a square circle). While the world as we experience it is misleading, its appearances are intersubjective (shared) in a conventional sense, and we could not navigate our empirical world without them.[114] By contrast, a particular individual may envisage something in front of them with no corresponding appearance in our

109 Dalal, "Śaṅkara," § 2.1.

110 Gupta, *An Introduction to Indian Philosophies*, 229.

111 King, *Indian Philosophy*, 217.

112 Dalal, "Śaṅkara," § 2.3.

113 Dalal, "Śaṅkara," § 2–2.3.

114 Dalal, "Śaṅkara," § 2.2.

intersubjective empirical world.[115] As one scholar summarizes Śaṅkara's view: "The world is not a complete delusion, because delusions are imagined by the mind, and the world is not the product of the delusions of a single mind."[116] In assigning the world this sort of middle-tier position, Śaṅkara wishes to avoid the nihilistic overtones of denying the world any existence at all while reaffirming the thoroughgoing oneness of *brahman*.

Śaṅkara has his sights on more than discerning truth or acquiring knowledge for its own sake. As is often the case in Indian philosophies, he also aims to liberate us from causes of our suffering and restore us to a more blissful state.[117] A major cause of our suffering, he believes, is ignorance. The ignorance underlying our experience of the world as distinct from *brahman* conceals and distorts reality.[118] We give names to plural and changing things as if they were ultimate reality because we do not understand the true nature of *brahman*. Our ignorance and the mistaken appearances it provokes obscure pure consciousness and place us in bondage.[119] We fail to appreciate that we are not separate from the perfect, pure consciousness that is *brahman*.

Since we already are *brahman*, we do not attain liberation by changing the nature of our existence. Instead, liberation entails enhancement of our knowledge and understanding. The shift needed is epistemological, not metaphysical.[120] To illustrate, Śaṅkara imagines ten children regrouping after crossing a river to make sure everyone made it to the other side. The children worry when each of them tallies only nine. Then a passerby points out that they all simply failed to include themselves in the count. The tenth child was there all along.[121] Enlightenment, then, requires a shift in perspective. While ultimately this transformation occurs through immediate recognition, we can prepare for it. Śaṅkara aims to assist with that preparation in various ways, including helping people progress from more to less pernicious mistaken impressions, and proposing ascetic and psychological practices to foster calmness, control of one's senses, and loving surrender.[122]

Although Śaṅkara does not concentrate on the puzzle of existence, his philosophy of radical oneness has significant implications for it. The puzzle of existence, as we have explored it, tends to take for granted a world comprising multiple things—including individual persons—that interact with one another. Śaṅkara denies these assumptions. We are not separate from the

115 King, *Indian Philosophy*, 215, 219.
116 King, *Indian Philosophy*, 219.
117 Dandekar, "Vedānta," 9546.
118 Gupta, *An Introduction to Indian Philosophies*, 233.
119 Potter, *The Encyclopedia of Indian Philosophies*, 6; Gupta, *An Introduction to Indian Philosophies*, 231.
120 Gupta, *An Introduction to Indian Philosophies*, 238.
121 Dalal, "Śaṅkara," § 4.
122 Torwesten, *Vedanta*, 146; Potter, *The Encyclopedia of Indian Philosophies*, 7–8.

world, and the world is not separate from *brahman*. All is one. (As noted even, the numerical reference to "one" is misleading to the extent that it suggests there is anything outside of *brahman*, or any alternative to it.) While Śaṅkara acknowledges that people experience the world as a plurality of continuously changing things, this is an illusion to overcome. There is no world as we typically understand it. The unreality of all distinctions eliminates the possibility of explaining the existence of one thing as the product of another. The world is a dependent effect of *brahman* but only as mistaken appearances borne of our ignorance. Śaṅkara's philosophy makes no room for a creation narrative or an account of one fully real thing giving rise to another.[123]

Not only do Śaṅkara's ideas resist explaining the world's existence in anything like a familiar form, but his philosophy is not oriented to addressing the problem in the first place. His thought is geared toward the liberation from suffering that is achieved through recognition that our impressions of the world are fundamentally erroneous. The crucial knowledge does not give us a better grip on the world but enables us to let it go. We do not aspire to understand better how things work but to see that they are not real; only *brahman* is.[124] It is in this sense that Śaṅkara does not so much explain the world as explain it away.[125]

There are significant similarities and differences between Nāgārjuna's *Madhyamaka* Buddhism and Śaṅkara's Advaita Vedānta.[126] Both thinkers deny that there is a plurality of independently existing things. They also both believe that the world as we commonly experience and think of it does not reflect ultimate reality. Accordingly, while acknowledging the phenomenal world, both accord it a lower status relative to absolute or ultimate reality. Moreover, both aspire to liberation from suffering through the elimination of ignorance. As a result of their respective worldviews, neither thinker's ideas embrace the puzzle of existence as commonly presented.

At the same time, their descriptions of ultimate reality differ dramatically. Nāgārjuna claims that nothing has an intrinsic, stable essence or independent existence. By contrast, Śaṅkara claims that reality is one independently existing, unchanging whole whose essence is pure consciousness. Additionally, in describing the world as we experience it, Nāgārjuna stresses the sheer arbitrariness of human conventions. Śaṅkara criticizes *Madhyamaka* Buddhism as nihilistic for this very reason, instead characterizing the world as a dependent effect of *brahman*.

---

123 Rambachan, "The Advaita Worldview," 69; Gupta, *An Introduction to Indian Philosophies*, 233.
124 King, *Indian Philosophy*, 216.
125 Torwesten, *Vedanta*, 123.
126 See Fowler, *Perspectives on Reality*, 242.

Each of the previous five chapters examines a distinctive approach to answering the question of why the world exists. However, not everyone accepts the invitation to offer an answer. In this chapter, we have considered thinkers whose ideas illustrate potential reasons for dismissing the puzzle of existence rather than proffering an answer. Hume, Edwards, and Russell attack the puzzle head on, advancing metaphysical views that make the question illegitimate in the form that others often frame it. They contend that it is legitimate to ask about why individual objects exist, but not to seek an explanation for the totality of all objects. Their view of what counts as an object does not include the sum or totality of numerous things. Nāgārjuna and Śaṅkara, while not as invested in explicitly engaging the topic, endorse worldviews that make the puzzle of existence unanswerable as commonly posed or simply beside the point.

Even in the case of thinkers who dismiss the topic, as we have seen, asking why the world exists provides an enlightening prism through which to encounter various schools of thought. In the next chapter, we turn to reasons for embracing the question, focusing on ways in which the quest for an answer sheds light on the nature of the problem and our own deepest commitments.

Questions for reflection and discussion

1. Does a collection of things have an existence above and beyond its individual members?
2. Is an infinite regress of explanations sufficient without an explanation of the whole?
3. What is the strongest argument that seeking a reason for the world's existence is illegitimate?
4. Is the idea of a necessarily existent being coherent?
5. Is it legitimate to think of the entire universe as an individual object?
6. Can one thing be responsible for another's existence apart from causal mechanisms?
7. How convincing is Nāgārjuna's argument that all things are interdependent?
8. Is Nāgārjuna right that there is no essential self underlying our conscious experience?
9. What is the most persuasive element of Śaṅkara's view that all being is one?
10. Can we reduce suffering by liberating ourselves from the most harmful illusions?

9.

Embracing the Inquiry

This concluding chapter takes stock of our exploration into the puzzle of existence. After noting why some might doubt the subject's value, we discuss gains from persisting in the inquiry. Notwithstanding limitations and uncertainties in our current understandings, the pursuit sheds light on the nature of the problem itself and our most deeply held views.

The last chapter considered arguments aimed at showing the illegitimacy of asking why the world exists. This author would not have embarked on the present study without deeming the opposing arguments more convincing. Nevertheless, we may acknowledge that even one who accepts the question's legitimacy might doubt the value of seeking an answer.

For one thing, success may not seem close at hand. We have examined several prominent approaches to explaining the world's existence: scientific (Chapter Three), mathematical (Chapter Four), axiological (Chapter Five), theistic (Chapter Six), and nomological (Chapter Seven). Each of these approaches faces unresolved challenges. To illustrate with some notable examples, a scientific approach risks falling into circularity by relying on methods that assume physical existence right from the start. A mathematical approach must derive the apparently dynamic experiential richness of our lives from the fixed, abstract relations of arithmetic, geometry, and other branches of the discipline. In addition to various challenges facing axiological and theistic approaches respectively, both confront the daunting task of reconciling pervasive suffering with a world grounded in goodness (whether in goodness itself, as in axiology, or in an all-good Creator god, as in theism). Lastly, while each nomological theory encounters distinctive objections of its own, they share the burden of explaining why we should accept that the very foundation of the world's existence is to be found in freestanding principles whose specific role is determining what shall exist. One consideration that may discourage investigation, then, is that no particular theory at present seems close to completing the task of explaining existence.

Another potentially discouraging consideration is the lack of any guarantee that the question has an answer. Russell could be right that the universe "is just there, and that's all."[1] Some explanations for the world posit a foundational brute fact that is responsible for its existence. But perhaps the foundational brute fact is the existence of the universe itself. There may simply be no answer to the question of why the world exists.

Moreover, even if there is an explanation for the world's existence, there is no guarantee that we can attain it. A wide range of factors might block our access. Some potential impediments concern human capacities. Critical data might be undetectable by our bodily senses, for example, or indispensable ideas might lie beyond our cognitive abilities. Another possibility is that important information is not available from our particular vantage point because its source is too distant or obscured by astronomical phenomena. Nor can we rule out the possibility of obstacles that are built into the fabric of the cosmos itself. For instance, obtaining an answer might hinge on computations that are impossible to complete in a finite amount of time.[2]

Why continue the quest in the face of such uncertainty? First, uncertainty cuts both ways. We cannot rule out the possibility that there *is* an answer to the puzzle of existence that is open to our understanding in principle, if not yet in fact. The limitations on our current knowledge include the inability to predict advances that might reduce our future limitations. Such advances could come in many forms, including measurement devices that enhance our observational capacities; new theories that further our understandings of the natural world; expansion in our ability to harness computational tools and artificial intelligence in our investigations; and transformations in the conceptual prisms through which we view the world. One era's most basic assumptions may be quaint errors noted in historical narratives recounted by later generations. Perhaps presently unforeseeable breakthroughs will yield insights pertinent to the puzzle of existence. Appropriate humility must not only acknowledge potential obstacles but also the chance that transformative developments will make us capable of more than we can presently envisage.

1 Russell, *Why I Am Not a Christian*, 134.

2 John Barrow, *Impossibility: The Limits of Science and the Science of Limits* (Oxford University Press, 1998), 210; A.K. Dewdney, *Beyond Reason: 8 Great Problems That Reveal the Limits of Science* (Wiley, 2004), 164–65.

Better Understanding the Problem

One motivation to continue the quest comes in the form of better understanding the problem. It is one thing to gesture toward some starting point or another for explaining the world's existence, and quite another to work through how components of a detailed account hang together. Attempts to explain the world's existence shed light on the strengths and vulnerabilities of the various competing approaches. The theories that we have examined make a significant contribution by elaborating how different forms of explanation could serve as the basis of an answer to the puzzle of existence. Advances within different disciplines and theoretical frameworks give us a better idea of what more fully developed solutions to the puzzle of existence might look like.

For example, contemporary physics introduces ideas about the natural world that people in earlier times could scarcely have imagined. While other ambitions underlie the discoveries, theorists like Edward Tryon view them as relevant to explaining the world's provenance. The most recent attempts to address the puzzle of existence from a scientific standpoint include exciting, revolutionary claims, such as the spontaneous, random appearance of matter in a vacuum.[3]

Similarly, present day mathematical answers to the puzzle of existence incorporate sophisticated developments unavailable to Pythagoras and others in earlier eras. Thinkers like Max Tegmark draw on cutting edge ideas from physics, mathematics, and computer science to show how we can understand the world's very foundation in terms of abstract mathematical relationships. Although we do not have a complete theory of existence based in mathematics—any more than we have a complete theory based in other sources—Tegmark's proposals for directions forward suggest the contours of theories to be further expounded in coming years. In this vein, Tegmark contends that symmetries—which play a critical role in both mathematics and physics—are a promising a basis for connecting the properties of mathematical relationships with the phenomena of our lived experience.[4]

Contemporary thinkers rooting existence in goodness, too, offer greater insights into the shape that refined axiological accounts can take. Millennia ago, Plato championed the idea that the good—with a real existence independent of the material world—has tremendous significance. Plato's good gives form to the material world and serves as an ideal standard toward which the virtuous aspire. However, like ancient Greek thinkers generally,

---

3 See Chapter Three.

4 See Chapter Four.

Plato assumes that there has always been some kind of physical stuff.[5] He wants to know why the cosmos is the way it is, but does not focus on why anything physical exists at all. Inspired by Plato, present day axiological thinkers like John Leslie contend that goodness explains not only the world's character but also its very existence. While Leslie would be the first to admit that his theory of "ethical requiredness" is a work in progress, it nevertheless gives substance to an axiological theory. Among other things, the account defends goodness as the foundational explanation of existence, and spells out the kinds of truths that obtain independently of physical existence, the relation that truths about value bear to particular things, and the sense in which goodness is responsible for the world's existence.[6]

To recall another example, Richard Swinburne advances strikingly contemporary arguments to defend a theistic solution for the puzzle of existence. With criteria for assessing theoretical claims modeled on those commonly used in the sciences, Swinburne makes the case that theism scores better than science itself as a basis for explaining the existence and nature of the physical world. We have also seen how a variety of contemporary nomological philosophers propose innovative solutions for the puzzle of existence rooted in laws or principles that are prior to the world and bear direct responsibility for its existence. Thus, Quentin Smith identifies the "Law of the Simplest Beginning" as the ultimate explanation for the world's existence. According to Smith's theory, this law—requiring that the simplest thing comes into existence in the simplest way—is responsible for the Big Bang singularity, which randomly generated the world around us in all its spectacular particularity. Smith's account largely centers on Big Bang cosmology while nevertheless grounding the world's existence in a nomological law that is prior to and independent of physics. Accordingly, it highlights the broad influence that the tremendous advances in contemporary science have had even on theorists who do not find an ultimate explanation for the world's existence in scientific theories.

In elaborating crucial elements of their theories, the thinkers we have examined illuminate the particular strengths and vulnerabilities of the competing general approaches to resolving the puzzle of existence. By studying them we better understand critical ways in which the approaches differ from one another. Their distinctive natures come into focus. Through examining a wide range of approaches in conjunction with one another we can also appreciate challenges or deep tensions that inevitably confront all attempts to explain the world's existence. Recognizing these challenges is important for at least two reasons. First, it may lead to advances through

5 See Chapter Two.

6 See Chapter Five.

better understandings of the structure and possible resolutions of shared challenges. Second, it ensures that we do not downgrade any particular theory of existence simply on the grounds of a difficulty that, in fact, every theory of existence encounters. We may profitably compare theories on their differing manners of addressing shared problems, while gleaning insights into the general nature of the challenges themselves.

Although there are multiple such overarching difficulties, the remainder of this section notes a particularly vexing one, which for convenience we will call the "problem of distinctness and interrelation."[7] The difficulty revolves around the relation between the physical world and whatever a theory identifies as its basis. More specifically, the problem concerns the following question: How similar to the physical world is the reason for its existence?

Let us consider what makes addressing this question potentially problematic. On the one hand, we want the world to be clearly dissimilar from whatever is responsible for its existence. After all, if the basis of the world's existence resembles the world itself, then we must worry about whether our understanding is circular. Beginning an account with something that is much like the world may seem tantamount to simply assuming from the start that which we had hoped to explain. Avoiding that kind of circular reasoning would mean establishing sufficient separation, dissimilarity, or "distinctness," between the world and the nature of that which explains its existence. From the vantage point of satisfying the distinctness requirement, the more *unlike* the ultimate explanation is to our everyday experience of the world, the better. We want an explanation appealing to something that is unmistakably independent of the things around us.

On the other hand, a satisfying theory must also allow for a connection between the world and whatever is responsible for its existence. Presumably, for X to explain Y, there must be some kind of linkage between them. If two things are utterly dissimilar, then we may worry that there is not a sufficient basis of "interrelation" between them for one to explain the other. From the

7 Another example of an overarching challenge concerns whether the explanation for the physical world's existence is ultimately rooted in something with a necessary or contingent existence. From one point of view, it may seem that the foundation must be necessary if it is to count as an explanation at all. One may worry that if we begin our narrative with something that could just as easily have not existed at all—and for whose existence we cannot identify a reason—then we have not really even gotten the task of explanation off the ground. On the other hand, if the foundation of all being is necessary in character, this raises the question of whether everything that flows or results from that foundation must be equally necessary. This implication would be problematic for anyone who believes that there are at least some things in the universe that are not thoroughly necessary, such as the making of choices by human beings or the behavior of subatomic particles under some interpretations of quantum physics. We have seen a variety of responses to this set of concerns. Tegmark, for example, explicitly advocates the view that the cosmos is rooted in necessity through and through, and embraces the conclusion that nothing escapes necessity's grip. Theists like Swinburne, and nomological thinkers like Parfit and Smith, on the other hand, identify contingent foundations, while insisting that they nevertheless count as genuine explanations for the world's existence.

vantage point of satisfying the interrelation requirement, the more *like* the ultimate explanation is to our everyday experience of the world, the better. We want an explanation appealing to something that could plausibly bring about physical states of affairs. The characteristics of distinctness and interrelation are potentially in tension, because the former tends to push the basis of existence away from the world, while the latter pulls it closer.[8] The two appealing features are not easy to combine because heightening one undercuts the other. The more convincingly an account fulfills the requirement of distinctness, the greater the challenge it encounters in showing how something so disparate from the things around us could be their reason for being.

The tension's ancient roots—dating to the dawn of Western philosophy—suggest its fundamentality as an obstacle for explanation. The problem of distinctness and interrelation is evident in the Milesians' quest for the *arche*: the world's first principle. The Milesians, like the ancients generally, take for granted that some kind of material stuff has always existed. Consequently, they do not ask—as we do—why the world exists at all. The Milesians do, however, pursue foundational understanding in seeking an explanation for the nature of the cosmos. Thales, the first known Western philosopher, suggests water as the *arche* (or first principle). The proposal has the advantage of familiarity. We ourselves consist largely of water, depend on it for life, and see it around us in multiple forms. Thus, Thales invites no worries that his *arche* is so alien or unknowable that it allows for no interconnection with the world.[9]

At the same time, while Thales' *arche* scores highly with respect to interrelation, it raises worries regarding distinctness. Can something so close to us—so obviously constituting part of our world—be up to the task of explaining its nature as a whole? Thales' younger compatriot Anaximander offers a theory that could not be more different. Anaximander's *arche*, which he calls the "boundless," is essentially undefined. Unlike the objects around us, Anaximander's boundless cannot be confined to a list of definite qualities. It is hard to imagine an *arche* that would score more highly with respect to distinctness. Yet, for that very reason, we may worry that it lacks sufficient interrelation with the cosmos whose nature it is supposed to explain.

The tension between distinctness and interrelation that already confronted the Milesians remains with us today, and is evident in the approaches that we have examined. Consider, for example, a scientific

8 Though in the course of different kinds of projects and using different terms to express the point, both Nicholas Rescher and Edward C. Halper capture the difficulties presented by this tension for those interested in identifying an ultimate explanation. See Rescher, *The Riddle of Existence*, 6–7, 23; Edward C. Halper, "The Hegelian Option," in *The Ultimate Why Question: Why Is There Nothing at All Rather Than Nothing Whatsoever?*, ed. John F. Wippel (Catholic University of America Press, 2012), 175–77.

9 See Chapter Two.

approach to explaining the world's existence. As a means of investigating the natural world, science gains credence from its reliance on empirical observations of physical objects. The discipline's methodologies are unparalleled in their usefulness for probing the behavior of electrons, trees, galaxies, rocks, and other objects in the world. An account explaining the world's existence through appeal to scientific theories scores highly on the interrelation criterion. The objects and concepts of science are manifestly close to us in nature. Consequently, they do not raise alarms of being too dissimilar from the world to have any kind of meaningful link with it. For the same reasons, however, a scientific approach scores poorly on the criterion of distinctness. Science's intrinsic reliance on existing physical properties as a basis of explanation prompts worries that it cannot provide a reason for the world's existence that escapes circularity.

Other approaches emphasize a radical contrast between the world and the ultimate source of its existence. Axiological accounts posit that goodness itself explains the world's existence. Recall that the axiological tradition has roots in the thought of Plato, who views the good as a perfect, unchanging, immaterial form that is fundamentally different in character from the flawed, fluctuating, material particulars in the world of our sensory experience.[10] While Plato does not directly confront the puzzle of existence, contemporary axiological theorists like John Leslie explicitly identify goodness as the answer to the question of why there is a physical world.[11] In defending his "theory of ethical requiredness," Leslie stresses that goodness has an existence that is prior to and independent of anything physical in character. Indeed, he specifically criticizes scientific explanations of the world's existence—like Edward Tryon's quantum vacuum theory—on the ground that they ultimately assume the very thing that they purport to explain. For example, Tryon's quantum vacuum, Leslie charges, is too much a part of the physical world itself to avoid circularity.[12]

In highlighting the radical distinctness of goodness, however, axiological answers may raise concerns regarding the other horn of the dilemma: the requirement of interrelation. Can we adequately understand or make sense of how the good itself can give rise to the existence of physical things? Similar challenges confront theistic and nomological accounts, which also maintain that only a basis of the world's existence that is radically distinct from it can avoid circularity. Some wonder how an immaterial being or an abstract law or principle could possibly bear responsibility for the fact that particular physical objects come to exist at a particular place and time.

10 See Chapter Two.
11 See Chapter Five.
12 Leslie, *Immortality Defended*, 73. Tryon's proposal is discussed in Chapter Three.

Max Tegmark's Mathematical Universe introduces another layer of complexity to our understanding of the problem of distinctness and interrelation. Axiological, theistic, and nomological theories like the ones we have examined are dualistic in that they identify a basis for the world's existence that is itself non-physical in character. Tegmark's theory, however, adopts a different kind of approach by entirely collapsing the distinction between the nature of the world and its reason for being. The MUH does not identify a distinct kind of existence that is responsible for physical being. Instead, it draws an equivalence between mathematical structures and the various universes that comprise the cosmic multiverse. In Tegmark's view, the theory provides a satisfying reason for the world's existence, because it is based in mathematical truths, which are self-explanatory.[13] Since there is no separation between the world and its foundation, Tegmark does not face the interrelation problem in the usual form. Instead, he confronts another critical challenge. If the world just is mathematics, then the theory must explain how the abstract truths of mathematics can account for all the ways that we experience the world around us. The point is not that this and the other questions we have noted cannot be addressed in satisfying ways, but, rather, that the problem of distinctness and interrelation poses a challenge for *all* theories of existence. In recognizing overarching challenges of this kind, we may attain better understandings of differences between competing approaches, as well of the challenges themselves.

Better Understanding Ourselves

The gains from investigating the puzzle of existence that we have considered thus far concern better understanding the problem. Other potential gains from investigating the puzzle come from better understanding ourselves.

Wrestling with the problem prompts reflection on our most deeply held beliefs. The everyday questions that normally confront us concern affairs taking place within a world whose existence we take for granted. Our inquiry, however, refuses to accept the world as a given. It seeks an explanation for the existence of the very stage on which we conduct our lives and all other inquiries. The quest's foundational nature requires us to consult our judgments on the most basic issues and how they hang together.

In engaging such a fundamental question, we might wish to avoid making assumptions that would tilt us toward particular conclusions right from the start. Nevertheless, no matter how many assumptions we strip away in

13 See Chapter Four.

the name of open-mindedness, it is impossible to avoid relying on at least some elemental principles that we bring to the inquiry rather than deriving them from it. Without adopting initial views on certain issues, we would have no basis on which to commence the inquiry or render any judgments whatsoever. For example, our analysis cannot help but draw on presuppositions about how to recognize theories that are worthy of attention in the first place, and the standards that will govern our assessment.

The inquiry requires not only that we draw on beliefs about basic issues but also that we make difficult choices about those beliefs, potentially leading us to doubt and revise them. People often find appealing a number of individual beliefs that do not fit comfortably together. For example, one might initially hold all of the following views: (1) that the reason for the world's existence is necessary in character; (2) that everything which exists in virtue of something necessary is itself necessary; and (3) that some things in the world are contingent (that is, they might have turned out differently). Notwithstanding a strong attraction to each of these views, one will find that asserting all three entails a contradiction.

The possibility that individually attractive views may be mutually inconsistent can make rumination on philosophical questions feel like packing everything for a big trip into a small suitcase. Stuffing boots into one side of the suitcase causes a jacket to pop out on the other. An accommodation here produces a crisis there. The luggage will not fit everything we wanted to take with us. Similarly, it may prove difficult to fit all of the ideas that one finds tempting into a single worldview. As a result, reckoning with the puzzle of existence compels us to assess which beliefs are the most indispensable.[14]

As we have seen in examining various theories, the puzzle of existence intersects with questions of foundational significance, such as how many fundamentally distinct kinds of objects there are. In examining a mathematical approach, for example, we discussed debates about whether mathematical objects have an existence that is independent of either physical things or the ideas in our own minds.[15] We considered a similar debate regarding the ontological status of goodness in our discussion of an axiological approach.[16] Another basic issue, discussed in the last chapter, concerns the standards for recognizing a question as meaningful or worthwhile.

Since extensive treatment here of all the questions that intersect with the puzzle of existence is infeasible, we will instead briefly highlight two

14 David Lewis makes similar points in discussing the nature of philosophical discourse. Speaking of the aim to bring one's opinions into equilibrium, Lewis notes that this involves distinguishing "the unacceptably counterintuitive consequences [from the] acceptably counterintuitive ones," and deciding "which prices are worth paying." David Lewis, *Philosophical Papers, Vol. I* (Oxford University Press, 1983), x.

15 See Chapter Four.

16 See Chapter Five.

particularly elemental issues that our exploration raises: whether there is an explanation for everything, and what we should regard as the most reliable guides to truth. The first of these issues concerns the extent to which we should expect there to be explanations for states of affairs (or the way things are). People are inherently inquisitive. The image of the insatiably curious child whose impulse to ask "Why" questions has no off switch is iconic for good reason. The practice of asking questions remains pervasive in our adult lives. Why is the traffic worse than usual today? Why is it raining so much lately? Why is a friend so happy today? What Aristotle called the desire of all people to know[17] has also fueled the development of intellectual disciplines that systematically investigate a wide range of fields from physics and chemistry to history and sociology. Notwithstanding the differences in these fields, they share a central focus on pursuing explanations. Together, these disciplines express the human desire to make sense of the world. Beyond surviving and reproducing, our species also yearns to understand.

It is an assumption underlying our posing of questions that they have answers. But is the assumption warranted? Might there not be brute facts: states of affairs that lack any explanation? After all, to *want* an explanation is not the same as having a reason to *expect* one.

In the most practical, immediate sense, we can ask whether to expect reasons for particular matters within distinct disciplines. Physicists debate whether there are reasons for subatomic events, for example. More broadly, we can ask whether the world is rationally ordered in a way that is open to our understanding. Do we have reasons to believe in reasons?

These questions have profound implications for the puzzle of existence. If there are explanations for all states of affairs—and in particular for all existing things—then we can expect an answer to the question of why the world exists. Indeed, some thinkers have built the assumption of a reason for the world's existence into the arguments for their particular theories. Examples include Leibniz's arguments for a theistic account[18] and Leslie's arguments for an axiological theory.[19] On the other hand, dropping the assumption opens the possibility that the world's existence simply lacks any underlying rationale.

A highly influential view denies that there are brute facts. In this vein, Leibniz maintains that there is a reason for everything,[20] calling this position the "principle of sufficient reason" (or "PSR").[21] In his words, "nothing

17 Aristotle, *Metaphysics*, trans. C.D.C. Reeve (Hackett, 2016), 1.980a22.
18 See Chapter Six.
19 See Chapter Five.
20 Leibniz, "Principles of Nature and Grace," § 7.
21 See Chapter Six.

happens without its being possible for him who should sufficiently understand things, to give a reason sufficient to determine why it is so and not otherwise."[22] Leibniz and other thinkers have expressed the principle in a number of ways, including that all events have causes, that there is an explanation for the existence of every contingent thing,[23] or, as van Inwagen puts it, "for every truth, for everything that is so, there is a sufficient reason for its being true or being so."[24] While Leibniz is perhaps the PSR's most famous proponent, something like it is evident as early as the work of ancient Greek philosophers like Anaximander and Parmenides.[25]

The PSR's advocates make a variety of points in its defense. Some claim that the PSR is self-evident. In this view, the PSR is so "obvious [and] intuitively clear" that one can grasp its veracity without "need of argumentative support."[26] Others maintain that the PSR is indispensable to other beliefs taken to be unassailable. Many assume, for example, that the world is intelligible. They accept as a matter of first principle that the world is rationally ordered and, thus, comprehensible. However, according to those advancing this argument, that assumption is inconsistent with the notion that reality lacks explanations for being as it is. Since the existence of explanations is intertwined with rationality and understanding, to deny that there are reasons is to deny the world's intelligibility.

A related argument claims that the PSR is indispensable to science, whose significance and legitimacy contemporary society largely takes for granted. What is science if not a quest for explanations? When physicists posit and test hypotheses, they are probing the causes for observable events. Scientific theories aim to identify general principles that help us to understand why events and interactions occur in the way that they do. Discarding the PSR, its proponents maintain, would undermine science, an enterprise that we can hardly imagine our lives without.[27] More broadly, intellectual inquiry revolves around the quest for explanations.[28]

Nevertheless, not everyone embraces the PSR. The PSR's opponents suggest that some things may not have explanations. There are brute facts, or, at least, we have no justification for assuming the contrary. In rebutting the claim that the PSR is self-evident, its adversaries observe that many people

22 Leibniz, "Principles of Nature and Grace," § 7.

23 Dahlstrom, "Being and Being Grounded," 128.

24 Peter van Inwagen, *Metaphysics*, 3rd ed. (Westview Press, 2008), 145.

25 Yitzhak Y. Melamed and Martin Lin, "Principle of Sufficient Reason," in *The Stanford Encyclopedia of Philosophy*, ed. Edward N. Zalta and Uri Nodelman (2023), § 5.3, accessed December 10, 2024, https://plato.stanford.edu/archives/sum2023/entries/sufficient-reason/.

26 Alexander R. Pruss, *The Principle of Sufficient Reason: A Reassessment* (Cambridge University Press, 2006), 189.

27 Pruss, *The Principle of Sufficient Reason*, 189.

28 Clack and Clack, *The Philosophy of Religion*, 36.

not only do not find the principle obvious or intuitive, but do not even find it persuasive or plausible.[29] The case against the PSR stresses that its defenders fail to furnish arguments establishing its veracity. That is, there is no positive basis for adopting it. We have no reason to accept the PSR as a truth of logic, for example, because that would mean that its denial entails a contradiction, which is not the case.[30] Empirical evidence also cannot demonstrate the PSR, since "[n]o amount of sensory experience will confirm it."[31]

While the PSR's supporters say that the principle is a precondition for intelligibility, its opponents note that our desire for intelligibility imposes no obligation on the universe to provide it. As one scholar writes, adopting the PSR "expresses a demand that things should be intelligible through and through ... [but] there is nothing that justifies this demand."[32] Similarly, another pair of scholars allow that "it may be nice and intellectually satisfying to believe that the universe is intelligible," but lament that it is "perfectly and disturbingly possible that life and the world are ultimately unintelligible."[33]

As for the argument that the PSR is indispensable to science, its opponents deny this as well. Even if science seeks explanations, they contend, this does not mean that the discipline depends on the assumption that *all* phenomena have explanations. Scientific inquiry may begin with brute facts and recognize that some events lack explanations while aiming to explain as much as possible.[34]

We have noted that the PSR's supporters present arguments aiming to reveal intolerable implications of *rejecting* the principle (such as the world's unintelligibility). The principle's opponents use a similar strategy in contending that *accepting* the principle entails an intolerable implication: the view that all events are determined in advance and can only unfold in one way. Those making this argument maintain that if there is a reason for everything and a full explanation for all events, then there is no room for contingency, probability, or randomness, which contradicts experience.

Of the subjects that intersect with the puzzle of existence, it is hard to imagine one more fundamental than the PSR. The principle operates more as a premise of inquiry than as the fruits of an investigation or proof. The nature of the arguments on both sides attest to its elemental status. Supporting the PSR seems already to presume the significance of providing reasons for our beliefs. At the same time, in emphasizing the lack of

29 Reichenbach, "Cosmological Argument."
30 Davies, *An Introduction to the Philosophy of Religion*, 61.
31 Davies, *An Introduction to the Philosophy of Religion*, 61.
32 Mackie, *The Miracle of Theism*, 85.
33 Clack and Clack, *The Philosophy of Religion*, 36–37.
34 Reichenbach, "Cosmological Argument."

justifications for adopting the PSR, the case against the principle inevitably relies on it. As one scholar notes, "even those who critique the PSR ... invoke it when they suggest that defenders of the principle have failed to provide a sufficient reason for thinking it is true."[35] To be sure, observing how pervasively our discourse and practices presume the PSR does not mean that reality must march to its requirements. Nevertheless, it does indicate that controversy over the PSR calls us to reflect on our most bedrock commitments.

A second elemental set of issues concerns what we should regard as the most reliable guides to truth. All of us have beliefs about what is (and is not) true. But how should we decide what to believe? In choosing among views, what should we trust? Asking why the world exists provokes reflection on these questions since we cannot help but rely on some kind of standards for assessing proposed answers. The puzzle of existence is so fundamental that it pushes us to consider the grounds of our most basic beliefs.

One question of this kind concerns the extent to which we should rely on common sense as a guide to truth. When presented with propositions about how the world is (or might be), we often have strong initial reactions about their plausibility (or likelihood of being true). Some claims may immediately strike us as easy to accept while others seem too farfetched to take seriously. These reactions often rest not on extensive intellectual analysis but, rather, on everyday assumptions that we carry around with us. We often do not make these assumptions explicit and may not always be attentive to the role that they play in our beliefs. In some contexts we refer to them as "gut reactions" or intuitive responses.

In reflecting upon such unreflective responses, how should we regard them? Should we heed them as valuable signposts or expose them as baseless feelings that distract us from sound reasoning? Suppose that we encounter a theory of why the world exists that initially strikes us as strange or bizarre. Should we treat that initial reaction as disqualifying the theory? Or should we be prepared to treat at least some claims that sound very weird as viable candidates for truth?[36] There are powerful considerations on both sides.

On the one hand, common sense has built up a store of credibility through its indispensable role in our daily lives. We could hardly navigate our environs safely and efficiently without it. The role of common-sense beliefs is so pervasive across every corner of our mental lives that it may seem folly to downgrade their significance.[37] The familiar expression, "It's

35 Reichenbach, "Cosmological Argument."

36 See Eric Schwitzgebel, "The Splintered Skeptic," in *Philosophy at 3AM: Questions and Answers with 25 Top Philosophers*, ed. Richard Marshall (Oxford University Press, 2014), 39.

37 Kit Fine, "Metaphysical Kit," in *Philosophy at 3AM*, 97–98.

just common sense"—when used to indicate the obvious truth of a proposition—reflects the importance that people often attach to it as a touchstone. The decision-making role that certain legal systems give to jurors drawing on ordinary experience—rather than specialized expertise—also reflects the weight that we often accord to common sense.[38]

On the other hand, the usefulness of common sense in some contexts might not guarantee its reliability as a guidepost to truth with respect to all kinds of questions. Consider the impact of natural selection on the way that we perceive the world. Evolutionary theory suggests that survival pressures equip us with perceptual abilities that make it easier to do things like find food and shelter while avoiding predators. The impressive track record of common sense in such contexts attests to its reliability regarding certain judgments. After all, we would probably not be here if it had too often led people astray regarding matters closely related to survival. At the same time, our activities are not limited to those with a direct bearing on survival. Human beings do not only eat, reproduce, and protect themselves from dangers but also sing, draw, build, create art, and, of most relevance here, ask profound questions regarding the nature of reality. These inquiries involve us in making judgments about contexts that are not critical to survival, such as those at the tiny scales of subatomic particles and the tremendous scale of galaxy clusters. There may be little reason to think that our common-sense beliefs serve as reliable guideposts to truth when we are considering claims about reality in these extraordinary settings.

Another question about the sources that we consult as guides to truth concerns the role of science. One can hardly doubt the usefulness of science as a means of investigating the behavior of physical things. The important role of science in aiming to understand the world is not controversial. However, a more contested question is whether we should view science as the *only* reliable guide to truth.

It is not hard to grasp why some might wish to treat science as the uniquely effective standard for evaluating claims. Science's emphasis on the empirical testing of hypotheses bespeaks an openness to correction through replicable observations. Other purported bases of truth may seem flimsy and ineffable in comparison. No other discipline can boast anything close to science's staggering record of tangible successes. Its theories make extraordinarily accurate predictions and foster technologies that transform our lives.

At the same time, there are reasons to question whether science deserves monopoly status as a standard of truth. Science explains events through theories that make generalizations about physical behavior. However, we cannot

38 See generally Norman J. Finkel and Bruce D. Sales, "Commonsense Justice: Old Roots, Germinant Ground, and New Shoots," *Psychology, Public Policy, and Law* 3 (1997): 227–38.

be certain that all truths take the form of claims about the interactions of physical entities. Humility cautions against assuming that any single discipline or methodology merits an exclusive place as our approach to inquiry.

Given our place on the historical timeline—in the midst of unprecedented technological progress catalyzed by the Scientific Revolution—we have an intense awareness of science's power as a basis for predicting and manipulating the things around us. We know as well the destruction made possible by the products of science, including nuclear weapons, attack drones, and poison gases. For all of the progress science generates, it also expands the scale of horrors that human beings can inflict on others. To be sure, the misuses of science do not necessarily bear on its reliability as a source of truth. Nevertheless, our reactions to these misuses suggest the importance of truths that may fall outside science's domain. Following the defeat of the Nazis, it seemed to some that little could be more certain than the immorality of genocide and other World War II atrocities.

Science can test claims about the dynamics of nuclear fusion, bomb damage zones, and the spread of fire in urban areas, but not the morality of human actions. More generally, many make claims—including in ethics, aesthetics, and spirituality—that may seem outside the scope of science. One response in the face of such claims is to conclude that they are not subjects of legitimate inquiry. From this standpoint, extra-scientific ideas might serve other uses—perhaps emotional or psychological—but we cannot legitimately assess their *truth*. A different response, however, is to recognize paths to truth beyond those that can be investigated through the methods of science. One may adopt such a view while also embracing science, as do many scientists and others. The controversy is not about the legitimacy of science, but, rather the exclusivity of science as a path to truth.

Science itself cannot resolve this dispute. One cannot demonstrate the exclusive legitimacy of science as an approach to inquiry from within science. More generally, the practice of science depends on principles that science cannot establish. Consider, for example, the conviction that everything subject to scientific inquiry is explicable through the interactions of physical things.[39] Scientists assume that differences in observed behaviors are the result of variations in physical processes, but science itself cannot substantiate that claim.[40] What empirical observations would confirm or disconfirm it?

Science also presumes that the physical laws describing the behavior of entities are the same in all times and places.[41] When observations contradict

39 Thomas Nagel, *Mind & Cosmos: Why the Materialist Neo-Darwinian Conception of Nature Is Almost Certainly False* (Oxford University Press, 2012), 4–5.

40 Enoch, *Taking Morality Seriously*, 135.

41 Hugh Rice, *God and Goodness* (Oxford University Press, 2003), 10.

expectations, researchers respond in a number of ways, including taking extra efforts to ensure the reliability of the observations or revising their theories. One option they do not consider is concluding that laws of nature change or differ from one laboratory to the next. Physicists take for granted that the behavioral regularities their theories describe apply across the full range of possible empirical observations.

The practice of science also depends on value judgments. Many assume that scientific objectivity requires methods that operate in a value-neutral manner.[42] However, practitioners cannot avoid relying on value judgments at a variety of stages in the practice of scientific inquiry. At the most general level, beliefs about what makes science valuable are critical to the development of its methodologies.[43] Arguments against alternative approaches to inquiry would be toothless without appeals to reasons sounding in considerations of value. Other aspects of science are equally dependent on value judgments, including choices about which topics to investigate. Characterizations of topics that merit time and attention—such as that they are significant, fascinating, or compelling—express value assessments.[44]

Value judgments also play a role in the development of scientific theories. In itself, the data that researchers collect does not determine the content of scientific theories, because there are multiple possible ways to theorize explanations for the same phenomena. As a result, researchers cannot help but rely on factors other than the observations themselves in forming theories. We have noted that science characteristically relies on a mode of reasoning known as abduction, or inference to the best explanation.[45] Abduction operates through the identification of criteria for assessing competing explanations. The selection of these criteria inevitably reflects judgments about what would make one theory *better than* another.[46]

In some cases—such as those with implications for the approval of pharmaceuticals or consumer products—these judgments may entail assigning relative weight to the potential human cost of error.[47] Even in cases not so clearly impacting public policy, practitioners make choices that inherently involve value judgments. Consider the preference that many scientists have for theories that are simple, beautiful, harmonious, or elegant. As physicist Steven Weinberg writes, "progress in physics is often guided by

42 David MacArthur, "Science and the Value of Objectivity," in *Facts and Values: The Ethics and Metaphysics of Normativity*, ed. Giancarlo Marchetti and Sarin Marchetti (Routledge, 2016), 233.

43 Richard Rudner, "The Scientist *Qua* Scientist Makes Value Judgments," *Philosophy of Science* 20, no. 1 (1963): 1–2.

44 Hilary Putnam, *Pragmatism and Realism*, ed. James Conant and Ursula M. Zeglen (Routledge, 2002), 17.

45 See Chapter Six.

46 Rudner, "The Scientist *Qua* Scientist Makes Value Judgments," 1–2.

47 Rudner, "The Scientist *Qua* Scientist Makes Value Judgments," 2–3.

judgments that can only be called aesthetic."[48] No laboratory experiment or set of observations can establish which traits to prefer. Nor can the normativity of scientific determinations be avoided by selecting the criteria according to their relative track records, since there is no value-free basis for assessing which past theories have been the most successful.[49]

Nothing has a more elemental impact on our understandings of the world than beliefs about the legitimate bases of belief. Decisions about what to trust shape all of our other views, as they inform our assessment of proposed explanations. Equally fundamental are questions about when to expect things in the world to have explanations in the first place. Engaging the puzzle of existence reveals aspects of our selves to ourselves by provoking reflection on foundational beliefs that inform attitudes and convictions that we often carry around without knowing or pausing to inquire why.

The Puzzle of Existence and the Quest for Meaning

Our exploration brings to light a deep interconnection between the quest for understanding and meaning. The history of art, literature, philosophy, science, religion—and so many other arenas of inquiry and expressions of culture—reflect a yearning to make sense of the world and to find or infuse our experience with meaning. We crave some kind of frame, overarching narrative, or set of ideas that allows us to regard our lives as more than a disconnected jumble of occurrences. We want to think of our lives as more than just one thing happening after another.

The puzzle of existence asks a "why" question, seeking an explanation for the world. The world is our home and the stage on which our lives play out. Thus, the reason for the world's being has potential implications for how we conceive of our place in it. The explanation for the world's existence has ramifications for the explanation of our own existence. There can hardly be a question more connected with the quest to make sense of our lives than "Why am I here?"

Any proposed reason for the world's existence prompts reflection on the implications for how we conceive of ourselves. These implications include potential answers to the kinds of questions that we have discussed in each of the chapters examining a particular approach to the puzzle of existence. What is the relation between us and the world's foundation? What is the subject of the most fundamental truths? What are the prospects for our

48 Weinberg, *Dreams of a Final Theory*, 17.

49 Hilary Putnam, *The Collapse of the Fact/Value Dichotomy, and Other Essays* (Harvard University Press, 2004), 32.

ability to grasp truths about the world? Does the world have any intrinsic purpose or meaning?

A particularly intriguing question concerns our own cosmic significance. Human beings share features that other things that we know about seem to lack. Among these is the drive to understand our lives as an ongoing narrative that aims toward ends that we consciously and intentionally adopt. This makes us wonder whether these sorts of ideas—like ends, narrative, and intention—have any place in the nature of the world's foundation? If so, does this suggest that the world was directed toward our existence from the beginning, or that we occupy a special place in the cosmos in some other way?

The interrelation between explanatory foundations and the cosmic significance of our lives has been manifest since people began pursuing both kinds of inquiry. Working only about a century after Thales—the earliest known Western philosopher—Pythagoras not only advanced beliefs about the foundational importance of mathematics, but combined them with beliefs about the progress of the human soul through reincarnation and numerous rituals governing the conduct of everyday life.[50] To note one other prominent example, Plato intertwined his positing of the good—the supreme immaterial form—with an account of the virtuous life as one aspiring toward an understanding of the good.[51] While many people today are most familiar with religion as a comprehensive source of foundational beliefs, all of the explanatory theories of the world's existence that we have examined entail implications for questions about the nature and place of human beings in the cosmos.

Insistent curiosity about the world and our relation to it is a fascinating aspect of our own nature, and itself one of the most mysterious things in the world that demands explanation. Grappling with these fundamental questions seems an essential part of what makes us human. It requires us to face—and assess—our most basic assumptions about reality, knowledge, and ourselves. Undertaking the inquiry promises a better understanding not only of the world but also of our own most deeply held beliefs.

50 See Chapter Four.

51 See Chapter Five.

Questions for reflection and discussion

1. What developments would contribute the most to our understanding of why the world exists?
2. Which answer to the puzzle of existence seems the most fully elaborated or plausible overall?
3. Could something quite different from the world around us be responsible for its existence?
4. What justifications can we give for assuming that there are reasons for the way things are?
5. How plausible is the possibility that the world's existence simply has no explanation?
6. If the world has no explanation, what does this mean for our ability to understand it?
7. Could methods of inquiry beyond science aid in our investigation into why the world exists?
8. How much do explanations for the world's existence shape our other fundamental beliefs?
9. Is there a common but erroneous assumption that throws us off in considering big questions?
10. How does thinking about the puzzle of existence relate to our quest for meaning?
11. What might we learn about ourselves in thinking about why anything exists?

Bibliography

Adamson, Peter. *Classical Philosophy: A History of Philosophy without Any Gaps*. Oxford University Press, 2014.

Alexander, H.G. *The Leibniz–Clarke Correspondence*. Manchester University Press, 1956.

Anglin, W.S., and J. Lambek. *The Heritage of Thales*. Springer, 1995.

Annas, Julia. *Platonic Ethics, Old and New*. Cornell University Press, 2018.

Anselm. *Monologion and Proslogion with the Replies of Gaunilo and Anselm*. Translated by Thomas Williams. Hackett, 1996.

Aquinas, Thomas. "The Treatise on the Divine Nature." Translated by Brian J. Shanley. In *Basic Works*, edited by Jeffrey Hause and Robert Pasnau. Hackett, 2014.

Aristotle. *Categories*. Translated by J.L. Ackrill. Clarendon Press, 1963.

——. *Generation of Animals*. Translated by A.L. Peck. Harvard University Press, 1942.

——. *Metaphysics*. Translated by C.D.C. Reeve. Hackett, 2016.

——. *Nicomachean Ethics*. Translated by Joe Sachs. Focus Publishing, 2002.

——. *Physics*, Books I–II. Translated by William Charlton. Clarendon Press, 1970.

Armour, Leslie. "Values, God, and the Problem about Why There Is Anything at All." *The Journal of Speculative Philosophy* 1, no. 2 (1987): 147–62.

Augustine, *City of God*. Translated by John O'Meara. Penguin Classics, 1984.

——. *Confessions*. Translated by R.S. Pine-Coffin. Penguin Classics, 1961.

Ayer, A.J. *Language, Truth, and Logic*. Dover, 1952.

Baggott, Jim. "What Einstein Meant by 'God Does Not Play Dice.'" *Aeon*, November 21, 2018. https://aeon.co/ideas/what-einstein-meant-by-god-does-not-play-dice.

Balaguer, Mark. "Fictionalism in the Philosophy of Mathematics." In *The Stanford Encyclopedia of Philosophy*, edited by Edward N. Zalta and Uri Nodelman. 2023. Accessed December 10, 2024. https://plato.stanford.edu/archives/spr2023/entries/fictionalism-mathematics/.

——. *Platonism and Anti-Platonism in Mathematics*. Oxford University Press, 1998.

Barrow, John D. *The Book of Nothing: Vacuums, Voids, and the Latest Ideas about the Origins of the Universe*. Pantheon Books, 2000.

——. *Impossibility: The Limits of Science and the Science of Limits*. Oxford University Press, 1998.

——. *New Theories of Everything*. Oxford University Press, 2007.

——. *Pi in the Sky: Counting, Thinking, and Being*. Clarendon Press, 1992.

——. *The Universe That Discovered Itself*. Oxford University Press, 2000.

Baum, Robert J. *Philosophy & Mathematics: From Plato to the Present*. Freeman, Cooper & Co., 1973.

Berkeley, George. *A Treatise concerning the Principles of Human Knowledge*. Edited by Kenneth Winkler. Hackett, 1982.

Broadie, Sarah. *Nature and Divinity in Plato's* Timaeus. Cambridge University Press, 2012.

Brown, Peter. *Augustine of Hippo: A Biography*. University of California Press, 1967.

Campbell, Joseph K. "Hume's Refutation of the Cosmological Argument." *International Journal for Philosophy of Religion* 40, no. 3 (1996): 159–73.

Carone, Gabriela Roxana. *Plato's Cosmology and Its Ethical Dimensions*. Cambridge University Press, 2005.

Carroll, Sean. *The Big Picture: On the Origins of Life, Meaning, and the Universe Itself*. Dutton, 2017.

Churchill, John. "Paradox and Identity in *Theology* by B.T. Herbert, and *Value and Existence* by John Leslie (Review)." *The Thomist: A Speculative Quarterly Review* 45, no. 4 (1981): 630–39.

Clack, Beverly, and Brian R. Clack. *The Philosophy of Religion: An Introduction*. 3rd ed. Polity, 2019.

Clay, Jenny Stray. *Hesiod's Cosmos*. Cambridge University Press, 2003.

Copleston, F.C. *A History of Medieval Philosophy*. Harper & Row, 1972.

Cornford, Francis MacDonald. *Plato's Cosmology: The* Timaeus *of Plato*. Hackett, 1997.

Craig, William Lane. *The Cosmological Argument from Plato to Leibniz*. Macmillan, 1980.

Curd, Patricia. "Parmenidean Monism." *Phronesis* 36, no. 3 (1991): 241–64.

Dahlstrom, Daniel O. "Being and Being Grounded." In *The Ultimate Why Question: Why Is There Nothing at All Rather Than Nothing Whatsoever?*, edited by John F. Wippel. Catholic University of America Press, 2012.

Dalal, Neil. "Śaṅkara." In *The Stanford Encyclopedia of Philosophy*, edited by Edward N. Zalta. 2021. Accessed December 22, 2024. https://plato.stanford.edu/archives/win2021/entries/shankara/.

Dandekar, R.N. "Vedānta." In *Encyclopedia of Religion*, edited by Lindsay Jones. Macmillan Reference, 2005.

Darwin, Charles. *The Origin of Species*. University of Pennsylvania Press, 1959.

——. *The Variation of Plants and Animals under Domestication*. John Murray, 1868.

Davies, Brian. *An Introduction to the Philosophy of Religion*. 4th ed. Oxford University Press, 2021.

——, ed. *Philosophy: A Guide to the Subject of Religion*. Georgetown University Press, 1998.

——. "The Problem of Evil." In *Philosophy of Religion: A Guide to the Subject*, edited by Brian Davies. Georgetown University Press, 2007.

Davies, Paul. *The Goldilocks Enigma: Why Is the Universe Just Right for Life?* First Mariner Books, 2008.

——. *The Mind of God: The Scientific Basis for a Rational World*. Simon & Schuster, 1992.

Davies, Paul, and J.R. Brown, eds. *The Ghost in the Atom: A Discussion of the Mysteries of Quantum Mechanics*. Cambridge University Press, 1986.

Dawkins, Richard. *The God Delusion*. Mariner Books, 2008.

Deltete, Robert. "Simplicity and Why the Universe Exists: A Reply to Quentin Smith." *Philosophy* 73, no. 285 (1998): 490–94.

Descartes, René. *Meditations on First Philosophy*. Translated by Michael Moriarty. Oxford University Press, 2008.

Deutsch, David. *The Fabric of Reality: The Science of Parallel Universes—and Its Implications*. Penguin Books, 1997.

Dewdney, A.K. *Beyond Reason: 8 Great Problems That Reveal the Limits of Science*. Wiley, 2004.

Dodds, Michael J. *The Unchanging God of Love*. 2nd ed. Catholic University Press, 2011.

Dorter, Kenneth. *Form and Good in Plato's Eleatic Dialogues:* The Parmenides, Theaetetus, Sophist, *and* Statesman. University of California Press, 1994.

Edwards, Paul. "The Cosmological Argument." In *Readings in the Philosophy of Religion*, edited by Baruch Brody. Prentice-Hall, 1974.

———. "The Cosmological Argument." In *The Cosmological Arguments: A Spectrum of Opinion*, edited by Donald R. Burrill. Anchor Books, 1967.

Enoch, David. *Taking Morality Seriously*. Oxford University Press, 2013.

Feldman, Fred. "The Open Question Argument: What It Isn't; and What It Is." *Philosophical Issues* 15 (2005): 22–43.

Ferris, Timothy. *The Whole Shebang: A State of the Universe(s) Report*. Simon & Schuster, 1997.

Feser, Edward. *Aquinas: A Beginner's Introduction*. Oneworld Publications, 2009.

Fine, Gail. *Plato on Knowledge and Forms: Selected Essays*. Oxford University Press, 2003.

Fine, Kit. "Metaphysical Kit." In *Philosophy at 3AM: Questions and Answers with 25 Top Philosophers*, edited by Richard Marshall. Oxford University Press, 2014.

Finkel, Norman J., and Bruce D. Sales. "Commonsense Justice: Old Roots, Germinant Ground, and New Shoots." *Psychology, Public Policy, and Law* 3 (1997): 227–41.

Fowler, Jeaneane. *Perspectives on Reality: An Introduction to the Philosophy of Hinduism*. Sussex Academic Press, 2002.

Friederich, Simon. "Fine-Tuning." In *The Stanford Encyclopedia of Philosophy*, edited by Edward N. Zalta and Uri Nodelman. 2023. Accessed December 10, 2024. https://plato.stanford.edu/archives/win2023/entries/fine-tuning/.

Gale, Richard M. *On the Nature and Existence of God*. Cambridge University Press, 1991.

Galilei, Galileo. "The Assayer." Translated by Stillman Drake. In *Discoveries and Opinions of Galileo*. Doubleday Anchor Books, 1957.

Garfield, Jay. *The Fundamental Wisdom of the Middle Way: Nagarjuna's Mulamadhyamakakarika*. Oxford University Press, 1995.

Gaunilo. "Reply on Behalf of the Fool." In Anselm, *Monologion and Proslogion with the Replies of Gaunilo and Anselm*, translated by Thomas Williams. Hackett, 1996.

Geach, P.T. "Aquinas." In *Three Philosophers*, edited by G.E.M. Anscombe and P.T. Geach. Basil Blackwell, 1963.

Goldschmidt, Tyron, ed. *The Puzzle of Existence: Why Is There Something Rather Than Nothing?* Routledge, 2013.

Gowers, Timothy. "Is Mathematics Discovered or Invented?" In *Meaning in Mathematics*, edited by John Polkinghorne. Oxford University Press, 2011.

Greene, Brian. *The Fabric of the Cosmos: Space, Time, and the Texture of Reality*. Vintage Books, 2004.

Gregory, Andrew. "Introduction." *Plato's Timaeus and Critias*. Translated by Robin Waterfield. Oxford University Press, 2008.

Gupta, Bina. *An Introduction to Indian Philosophies: Perspectives on Reality, Knowledge, and Freedom*. Routledge, 2012.

Guthrie, W.K.C. "Flux and Logos in Heraclitus." In *The Presocratics: A Collection of Critical Essays*, edited by Alexander P.D. Mourelatos. Princeton University Press, 1993.

——. *A History of Greek Philosophy, Vol. 2: The Presocratic Tradition from Parmenides to Democritus*. Cambridge University Press, 1965.

Halper, Edward C. "The Hegelian Option." In *The Ultimate Why Question: Why Is There Nothing at All Rather Than Nothing Whatsoever?*, edited by John F. Wippel. Catholic University of America Press, 2012.

Hamilton, Sue. *Indian Philosophy: A Very Short Introduction*. Oxford University Press, 2001.

Hamlin, Colin. "Towards a Theory of Universes: Structure Theory and the Mathematical Universe Hypothesis." *Synthese* 194 (2017): 571–91.

Hare, R.M. *The Language of Morals*. Oxford University Press, 1952.

Harvey, Peter. *An Introduction to Buddhism: Teachings, History and Practices*. Cambridge University Press, 1990.

Hawking, Stephen, and Leonard Mlodinow. *The Grand Design*. Bantam Books, 2010.

Hebblethwaite, Brian. "Review of *Universes* and *Physical Cosmology and Philosophy*, by John Leslie." *Zygon: A Journal of Religion and Science* 26, no. 2 (1991): 317–24.

Heidegger, Martin. "What Is Metaphysics?" In *Basic Writings from* Being and Time *to* The Task of Thinking, edited by David Farrell Krell. Harper & Row, 1977.

Heller, Michael. *Ultimate Explanations of the Universe*. Springer, 2009.

Helm, Paul. "Eternality." In *Philosophy: A Guide to the Subject of Religion*, edited by Brian Davies. Georgetown University Press, 1998.

——. "Review of *Value and Existence* by John Leslie." *The Philosophical Quarterly* 30, no. 121 (1980): 376–77.

Hesiod. *Theogony*. Translated by M.L. West. Oxford University Press, 1988.

Hick, John. *Evil and the God of Love*. Harper & Row, 1977.

Hoffman, J., and G.D. Rosenkrantz. *The Divine Attributes*. Blackwell, 2002.

Holder, Rodney. *God, the Multiverse, and Everything*. Routledge, 2016.

Hume, David. *Dialogues concerning Natural Religion*. 2nd ed. Edited by Norman Kemp Smith. Thomas Nelson and Sons, 1947.

Hussey, Edward. *The Presocratics*. Charles Scribner's Sons, 1972.

Jewish Publication Society, ed. *The Five Books of Moses*. 1999.

Kahn, Charles H. *Pythagoras and the Pythagoreans: A Brief History*. Hackett, 2001.

——. "Pythagorean Philosophy before Plato." In *The Pre-Socratics: A Collection of Critical Essays*, edited by Alexander P.D. Mourelatos. Princeton University Press, 1993.

Kane, G. Stanley. "The Failure of Soul-Making Theodicy." *International Journal for Philosophy of Religion* 6 (1975): 1–22.

Kant, Immanuel. *Critique of Pure Reason*. Translated by Norman Kemp Smith. St. Martin's Press, 1963.

Karamanolis, George. *The Philosophy of Early Christianity*. Routledge, 2013.

Kaye, Sharon M. *Medieval Philosophy*. Oneworld Publications, 2008.

Kenny, Anthony. *Medieval Philosophy*. Oxford University Press, 2005.

King, Richard. *Indian Philosophy: An Introduction to Hindu and Buddhist Thought*. Georgetown University Press, 1999.

Kline, Morris. *Mathematical Thought from Ancient to Modern Times*. Oxford University Press, 1992.

Koterski, Joseph W. *An Introduction to Medieval Philosophy*. Wiley-Blackwell, 2008.

Krauss, Lawrence. *A Universe from Nothing: Why There Is Something Rather Than Nothing*. Free Press, 2012.

Leibniz, Gottfried Wilhelm. "Principles of Nature of Grace and of Reason." Translated by Robert Latta and George R. Montgomery. In *Discourse on Method and Other Writings*, edited by Peter Loptson. Broadview Press, 2012.

Leslie, John. "A Cosmos Existing through Ethical Necessity." *Philo* 12, no. 2 (2009): 172–87.

——. *Immortality Defended*. Wiley-Blackwell, 2007.

——. *Infinite Minds: A Philosophical Cosmology*. Clarendon Press, 2001.

——. "Our Place in the Cosmos." *Philosophy* 75, no. 291 (2000): 5–24.

——. "A Proof of God's Reality." In *The Puzzle of Existence: Why Is There Something Rather Than Nothing?*, edited by Tyron Goldschmidt. Routledge, 2003.

——. "The Theory That the World Exists Because It Should." *American Philosophical Quarterly* 7, no. 4 (1970): 286–98.

——. *Value and Existence*. Rowman and Littlefield, 1979.

Lewis, David. *Philosophical Papers, Vol. I*. Oxford University Press, 1983.

Linnebo, Øystein. *Philosophy of Mathematics*. Princeton University Press, 2017.

Lowe, E.J. "Why Is There Anything at All?" *Proceedings of the Aristotelian Society* 70 (1996): 111–20.

Lyas, Colin. "Review of *Value and Existence* by John Leslie." *Philosophy* 55, no. 212 (1980): 275–77.

MacArthur, David. "Science and the Value of Objectivity." In *Facts and Values: The Ethics and Metaphysics of Normativity*, edited by Giancarlo Marchetti and Sarin Marchetti. Routledge, 2016.

MacDonald, Scott. *Being and Goodness: The Concept of the Good in Metaphysics and Philosophical Theology*. Cornell University Press, 1990.

Mackie, J.L. *Ethics: Inventing Right and Wrong*. Penguin Books, 1990.

——. "Evil and Omnipotence." *Mind* 64 (1955): 200–12.

——. *The Miracle of Theism*. Oxford University Press, 1982.

Maddy, Penelope. "How Applied Mathematics Became Pure." *The Review of Symbolic Logic* 1, no. 1 (2008): 16–41.

Malcolm, Norman. "Anselm's Ontological Arguments." *Philosophical Review* 69 (1960): 41–62.

Mann, William E. "The Perfect Island." *Mind* 85, no. 339 (1976): 417–21.

Manson, Neil, ed. *God and Design: The Teleological Argument and Modern Science*. Routledge, 2003.

Marcus, Russell, and Mark McEvoy. *An Historical Introduction to the Philosophy of Mathematics: A Reader*. Bloomsbury, 2016.

Mates, Benson. *The Philosophy of Leibniz: Metaphysics and Language*. Oxford University Press, 2003.

Mawson, T.J. *Belief in God: An Introduction to the Philosophy of Religion*. Clarendon Press, 2005.

Melamed, Yitzhak Y., and Martin Lin. "Principle of Sufficient Reason." In *The Stanford Encyclopedia of Philosophy*, edited by Edward N. Zalta and Uri Nodelman. 2023. Accessed December 10, 2024. https://plato.stanford.edu/archives/sum2023/entries/sufficient-reason/.

Mendel, Gregor Johann. "Experiments concerning Plant Hybrids." *Proceedings of the Natural History Society of Brünn* IV (1865): 3–47.

Mermin, N. David. "Is the Moon There When Nobody Looks? Reality and the Quantum Theory." *Physics Today* 4 (1985): 38–47.

Moore, G.E. *Principia Ethica*. Cambridge University Press, 1956.

Nagel, Thomas. *Mind & Cosmos: Why the Materialist Neo-Darwinian Conception of Nature Is Almost Certainly False*. Oxford University Press, 2012.

Nielsen, Kristian Hvidtfelt. "Surveying the Attitudes of Physicists concerning Foundational Issues of Quantum Mechanics." Accessed December 8, 2024. https://arxiv.org/pdf/1612.00676.pdf.

Norsen, Travis. *Foundations of Quantum Mechanics: An Exploration of the Physical Meaning of Quantum Theory*. Springer, 2017.

Nozick, Robert. *Philosophical Explanations*. Harvard University Press, 1981.

O'Connell, Robert J. *St. Augustine's Early Theory of Man, A.D. 386–391*. Harvard University Press, 1968.

Oppy, Graham. *Arguing about Gods*. Cambridge University Press, 2006.

——. "Ontological Arguments." In *The Stanford Encyclopedia of Philosophy*, edited by Edward N. Zalta and Uri Nodelman. 2023. Accessed December 10, 2024. https://plato.stanford.edu/archives/fall2023/entries/ontological-arguments/.

Paley, William. *Natural Theology: Or, Evidence of the Existence and Attributes of the Deity, Collected from the Appearances of Nature*. Cambridge University Press, 2009.

Parfit, Derek. "Why Anything? Why This? Part 1." *London Review of Books* 20, no. 2 (1998): 24–27.

——. "Why Anything? Why This? Part 2." *London Review of Books* 20, no. 3 (1998): 20–23.

Peretó, J., J.L. Bada, and A. Lazcano. "Charles Darwin and the Origin of Life." *Origins of Life and Evolution of Biospheres* 39, no. 5 (2009): 395–406.

Plantinga, Alvin. *God, Freedom, and Evil*. Harper & Row, 1974.

——. *The Nature of Necessity*. Oxford: Clarendon Press, 1974.

Plato. *Laws*. Translated by Tom Griffith. Cambridge University Press, 2016.

——. *Phaedo*. Translated by R.S. Bluck. Routledge, 2014.

——. *Republic*. Translated by Desmond Lee. Penguin Books, 1974.

——. *Symposium*. Translated by C.J. Rowe. Aris & Phillips, 1998.

——. *Timaeus*. Translated by Robin Waterfield. Oxford: Oxford University Press, 2008.

Potter, Karl, H. *The Encyclopedia of Indian Philosophies, Vol. 3: Advaita Vedanta Up to Samkara and His Pupils*. Princeton University Press, 2014.

Pradhan, Ramesh Chandra. *Metaphysical Idealism, a Contemporary Perspective*. Cambridge Scholars Publishing, 2023.

Pruss, Alexander R. *The Principle of Sufficient Reason: A Reassessment*. Cambridge University Press, 2006.

——. "A Restricted Principle of Sufficient Reason and the Cosmological Argument." *Religious Studies* 40 (2004): 165–79.

Puccetti, Roland. "Does the Universe Exist Because It Ought To? A Critique of Extreme Axiarchism." *Dialogue* 32, no. 4 (1993): 651–57.

Pucci, Pietro. *Hesiod and the Language of Poetry*. Johns Hopkins University Press, 1977.

Putnam, Hilary. *The Collapse of the Fact/Value Dichotomy, and Other Essays*. Harvard University Press, 2004.

——. *Pragmatism and Realism*. Edited by James Conant and Ursula M. Zeglen. Routledge, 2002.

Raatikainen, Panu. "Gödel's Incompleteness Theorems." In *The Stanford Encyclopedia of Philosophy*, edited by Edward N. Zalta. 2022. Accessed December 10, 2024. https://plato.stanford.edu/archives/spr2022/entries/goedel-incompleteness/.

Rambachan, Anantanand. *The Advaita Worldview: God, World, and Humanity*. State University of New York Press, 2006.

Ratzsch, Del. "Review of *Value and Existence* by John Leslie." *Noûs* 17, no. 1 (1983): 113–16.

Rees, Martin. *Before the Beginning: Our Universe and Others*. Basic Books, 1997.

——. *Just Six Numbers: The Deep Forces That Shape the Universe*. Basic Books, 2000.

Reichenbach, Bruce R. "Cosmological Argument." In *The Stanford Encyclopedia of Philosophy*, edited by Edward N. Zalta and Uri Nodelman. 2023. Accessed December 10, 2024. https://plato.stanford.edu/archives/win2023/entries/cosmological-argument/.

——. "Explanation and the Cosmological Argument." In *Contemporary Debates in Philosophy of Religion*, edited by Michael L. Peterson and Raymond J. VanArragon. Blackwell, 2004.

Rescher, Nicholas. *Axiogenesis: An Essay in Metaphysical Optimalism*. Lexington Books, 2010.

——. *The Riddle of Existence: An Essay in Idealistic Metaphysics*. University Press of America, 1984.

Rice, Hugh. *God and Goodness*. Oxford University Press, 2003.

Ross, W.D. *The Right and the Good*. Clarendon Press, 1930.

Rowe, Christopher. "The Form of the Good and the Good in Plato's *Republic*." In *Pursuing the Good: Ethics and Metaphysics in Plato's* Republic, edited by Douglas Cairns, Fritz-Gregor Herrmann, and Terry Penner. Edinburgh University Press, 2007.

Rubenstein, Mary-Jane. *Worlds without End: The Many Lives of the Multiverse*. Columbia University Press, 2014.

Rudner, Richard. "The Scientist *Qua* Scientist Makes Value Judgments." *Philosophy of Science* 20, no. 1 (1963): 1–6.

Russell, Bertrand. *Why I Am Not a Christian: And Other Essays on Religion and Related Subjects.* Routledge, 1957.

Rynhold, Daniel. *An Introduction to Medieval Jewish Philosophy*. I.B. Tauris, 2009.

Schlesinger, G. "The Problem of Evil and the Problem of Suffering." *American Philosophical Quarterly* 1, no. 3 (1964): 244–47.

Schopenhauer, Arthur. *On the Fourfold Root of Sufficient Reason*. Translated by Karl Hillebrand. George Bell and Sons, 1903.

Schwitzgebel, Eric. "The Splintered Skeptic." In *Philosophy at 3AM: Questions and Answers with 25 Top Philosophers*, edited by Richard Marshall. Oxford University Press, 2014.

Smith, Norman Kemp. "Introduction." In *Hume's Dialogues concerning Natural Religion*, edited by Norman Kemp Smith. Thomas Nelson and Sons, 1947.

Smith, Quentin. "Simplicity and Why the Universe Exists." *Philosophy* 72 (1997): 125–32.

Smolin, Lee. "Scientific Alternatives to the Anthropic Principle." In *Universe or Multiverse*, edited by Bernard Carr. Cambridge University Press, 2007.

Stenger, Victor. *The Fallacy of Fine-Tuning: Why the Universe Is Not Designed for Us*. Prometheus Books, 2011.

Stratton-Lake, Philip. "Introduction." In W.D. Ross, *The Right and the Good.* Clarendon Press, 1930.

Swinburne, Richard. *The Existence of God*. Clarendon Press, 2004.

——. *Is There a God?* Oxford University Press, 1996.

Taliaferro, Charles. "Personal." In *Philosophy: A Guide to the Subject of Religion*, edited by Brian Davies. Georgetown University Press, 1998.

Tarnas, Richard. *The Passion of the Western Mind: Understanding the Ideas That Have Shaped Our World View*. Harmony Books, 1991.

Tegmark, Max. "Is 'The Theory of Everything' Merely the Ultimate Ensemble Theory?" *Annals of Physics* 270, no. 1 (1998): 1–51.

——. "The Mathematical Universe." *Foundations of Physics* 38 (2008): 101–50.

——. *Our Mathematical Universe: My Quest for the Ultimate Nature of Reality*. Knopf, 2014.

——. "Shut up and Calculate." 2007. Accessed December 8, 2024. https://arxiv.org/abs/0709.4024.

Tegmark, Max, Piet Hut, and Mark Alford. "On Math, Matter and Mind." *Foundations of Physics* 36, no. 6 (2006): 765–94.

Te Velde, Rudi. *Aquinas on God: The 'Divine Science' of the* Summa Theologiae. Ashgate, 2006.

Torwesten, Hans. *Vedanta: Heart of Hinduism*. Adapted by Loly Rossett, from a translation from the German by John Phillips. Grove Weidenfeld, 1991.

Vallicella, William F. "Divine Simplicity." In *The Stanford Encyclopedia of Philosophy*, edited by Edward N. Zalta and Uri Nodelman. 2023. Accessed December 8, 2024. https://plato.stanford.edu/archives/win2023/entries/divine-simplicity/.

Van Inwagen, Peter. *Metaphysics*. 3rd ed. Westview Press, 2008.

——. "Why Is There Anything at All?" *Proceedings of the Aristotelian Society* 70 (1996): 95–110.

Wainwright, William J. "Review of *Value and Existence* by John Leslie." *The Philosophical Review* 90, no. 2 (1981): 318–21.

Weinberg, Steven. *Dreams of a Final Theory*. Pantheon Books, 1992.

——. *The First Three Minutes: A Modern View of the Origin of the Universe*. Basic Books, 1993.

Westerhoff, Jan Christoph. "Nāgārjuna." In *The Stanford Encyclopedia of Philosophy*, edited by Edward N. Zalta and Uri Nodelman. 2024. Accessed December 18, 2024. https://plato.stanford.edu/archives/sum2024/entries/nagarjuna/.

——. *Nāgārjuna's Madhyamaka: A Philosophical Introduction*. Oxford University Press, 2009.

White, Stephen. "Milesian Measures: Time, Space, and Matter." In *The Oxford Handbook of Presocratic Philosophy*, edited by Patricia Curd and Daniel W. Graham. Oxford University Press, 2008.

Wigner, Eugene. "The Unreasonable Effectiveness of Mathematics in the Natural Sciences." *Communications in Pure and Applied Mathematics* 13, no. 1 (1960): 1–14.

Wilson, Catherine. *The Invisible World: Early Modern Philosophy and the Invention of the Microscope*. Princeton University Press, 1995.

Witherall, Arthur. "The Fundamental Question." *Journal of Philosophical Research* 26 (2001): 53–87.

Young, Robert. "Review of *Value and Existence* by John Leslie." *Religious Studies* 17, no. 1 (1981): 129–31.

Index

About the Publisher

The word "broadview" expresses a good deal of the philosophy behind our company. Our focus is very much on the humanities and social sciences—especially literature, writing, and philosophy—but within these fields we are open to a broad range of academic approaches and political viewpoints. We strive in particular to produce high-quality, pedagogically useful books for higher education classrooms—anthologies, editions, sourcebooks, surveys of particular academic fields and sub-fields, and also course texts for subjects such as composition, business communication, and critical thinking. We welcome the perspectives of authors from marginalized and underrepresented groups, and we have a strong commitment to the environment. We publish English-language works and translations from many parts of the world, and our books are available world-wide; we also publish a select list of titles with a specifically Canadian emphasis.

broadview press

This book is made of paper from well-managed FSC® - certified forests, recycled materials, and other controlled sources.